Praise for
The Valachi Papers

"What Valachi did is beyond measure. Before him we had no concrete evidence that anything like the Cosa Nostra existed. In the past we've heard that so and so was a syndicate man, and that was about all. But Valachi named names. He revealed what the structure was and how it operates. In a word, he showed us the face of the enemy."
—WILLIAM HUNDLEY,
former chief, Organized Crime
and Racketeering Section,
U.S. Department of Justice

"A frightening book because it shows how easily a secret criminal organization can become a spreading cancer in an affluent society like ours."
—JOHN BARKHAM, *Saturday Review Syndicate*

"Exciting as well as informative, as fascinating as fiction, a bloody history of the Mafia as lived by one of its members."
—*New York Times Book Review*

"Shows that the tabloids' most lurid headlines are actually rank understatements of actuality."
—*Detroit Free Press*

Also by Peter Maas

FICTION

Father and Son
Made in America

NONFICTION

Manhunt
Marie: A True Story
King of the Gypsies
Serpico
The Rescuer
Underboss
The Terrible Hours
In a Child's Name

The Valachi Papers

Peter Maas

Perennial

An Imprint of HarperCollinsPublishers

First Perennial edition published 2003.

Designed by Jamie Kerner-Scott

Library of Congress Cataloging-in-Publication Data is available.

ISBN 0-06-050742-X

03 04 05 06 07 ❖/RRD 10 9 8 7 6 5 4 3 2

This is for John Michael

Author's Note

The Valachi Papers could not in fact have been written without the help and encouragement of a number of concerned individuals, particularly in the Department of Justice and the Federal Bureau of Investigation. Because of the sensitivity of the subject and the controversy surrounding it, I can only express my appreciation to them without singling them out.

Aside from those whose anonymity must be protected, I am especially grateful to Rosamond Dana, Irene DeBenham, Patricia Orr, and Pat Sloatman for their invaluable assistance in preparing and researching the manuscript. Finally my thanks to my editor, Arthur C. Fields, for his expert guidance in seeing the project through.

The Story Behind
"The Valachi Papers"

In late June 1964—nobody can recall the exact date—Joseph Valachi, who in the words of then Attorney General Robert F. Kennedy had provided the "biggest single intelligence breakthrough yet in combating organized crime and racketeering in the United States," was urged by the Department of Justice to write a history of his underworld career.

Valachi was the first person ever to admit belonging to or openly to talk about a huge, tightly knit, secret criminal conspiracy in this country, indeed an entire subculture of evil, popularly known as the Mafia until he revealed its real name, the Cosa Nostra. Valachi had already been interrogated by the Bureau of Narcotics and the Federal Bureau of Investigation and had appeared before a Senate investigating subcommittee chaired by Senator John L. McClellan of Arkansas. Because of his importance, he was now asked to set down on paper his own recollections, since some fragmentary information might turn up that had been overlooked in formal questioning. Usually such personal chronicles are fairly brief, cut-and-dried affairs. But Valachi has an extraordinary and precise sense of recall, and he literally accepted the Justice Department's exhortation that this was the only way he could expiate his criminal past and square his accounts with society. For thirteen months, wearing out a succession of ballpoint pens, he laboriously churned out more than 300,000 words on note pads, telling about his career in the Cosa Nostra. What gradually began to take shape was no mere recital of crimes, but a unique and exhaustive portrayal, virtually day by day over a thirty-year period, of life in the organized underworld. For someone whose schooling had stopped in the seventh grade, it was a monumental task, and to keep Valachi at work, the Justice Department held out the hope to him that when completed, his memoirs might be published, so that everyone finally would be able to see what one law enforcement officer had called "the face of the enemy."

Having originally broken the Valachi story and his revelations about the existence of the Cosa Nostra in an article in the *Saturday Evening Post*,* I was particularly anxious to see what he had to say. Subsequently, and unofficially, I obtained a copy of his manuscript to judge its potential. It had been typed by Justice Department personnel and totaled a staggering 1,180 pages. I was fascinated by what I found in them. While there have been books in the past about something variously and vaguely called the Mob, the Mafia, or the Syndicate, they have always been produced by outsiders—sociologists, district attorneys, crime re-

---

*Not that it would have remained under wraps forever. At the time my article appeared, some parts of the story, gathered from her own sources by reporter Miriam Ottenberg, ran in the *Washington Star*.

porters, ex-detectives, and so on. Valachi told the story in dispassionate detail for the first time as only an insider can tell it—often brutal, sometimes funny, always real.

Quite apart from the editing that the manuscript would require—Valachi's style made William Faulkner look like a simple subject-and-predicate man—there was another obstacle that stood in the way of its immediate publication: a Bureau of Prisons regulation which barred federal inmates from publishing anything about their crimes. For several months the Justice Department considered waiving this rule, while Valachi sat in his cell waiting for a decision. Finally, in December 1965, Attorney General Nicholas deB. Katzenbach, who had by then succeeded Kennedy, informed Valachi that he had authorized the manuscript's release. I was named, with Valachi's consent, to edit it. I was chosen because I had been the first to request the assignment, and in a "Dear Peter" letter, dated December 22, 1965, signed by the Justice Department's Director of Public Information Jack Rosenthal, I was officially notified of my selection.

This letter, in noting that the department's decision was "an exception" to its policy of not allowing prisoners to publish stories about their criminal careers, was quite specific about the reasons for the exception in Valachi's case:

1. He did not undertake writing the book for personal or selfish purposes, but at the instance of the Department, for the law enforcement purpose of seeking intelligent information beyond what he had recounted in interviews.

2. His notoriety is not the result merely of the dramatic nature of his crimes, or career in crime, but is rather the result of his

disclosures about the nature of organized crime—disclosures made in public at the instance of the Department and Senator McClellan.

3. It is possible and perhaps likely that law enforcement can be benefitted substantially by publication of the book.

The last paragraph especially reflected the attitude of the Organized Crime and Racketeering Section of the Justice Department. Valachi's televised appearances before a Senate subcommittee headed by Senator McClellan, of which more later, had not gone very well. There was still considerable controversy over whether an organization like the Cosa Nostra really existed. It was hoped that the sheer mass of data provided by Valachi would do much to clear this up, as well as to counteract a great deal of public indifference on the subject. There was also the hope that publication of the book would encourage other informers to come forward.

Arrangements were then made for me to work with Valachi in the District of Columbia Jail, where he was being kept under heavy guard. It was my impression that we would proceed without fanfare, and the book would just come out. I was mistaken. On December 27, to my astonishment, and presumably the Justice Department's, word of the project, and my part in it, was carried in newspapers and on radio and television across the country. I later discovered what had happened. The Justice Department had received a number of inquiries about the possibility of a Valachi book, and Rosenthal, when he wrote to me, also wrote to them about the Attorney General's decision. One of the recipients was the Associated Press, which promptly put the news out.

At the time, however, it did not seem terribly important, and on January 3, 1966, before my first meeting with Valachi, I conferred with William G. Hundley, Chief of the Organized Crime and Racketeering section, and Paul F. McArdle, who was President of the District of Columbia Bar Association and who at the request of the Justice Department had agreed to represent Valachi. We were there in Hundley's office to sign a Memo of Understanding that the department had prepared relative to the editing and publishing of Valachi's raw copy. It incorporated essentially the material in Rosenthal's letter, along with the right to "require the elimination of material the Department deems injurious to law enforcement purposes." I had already been informed verbally that this was in reference to harsh charges of corruption that Valachi had leveled in his manuscript against some members of the Bureau of Prisons and dubious conduct against others in the Bureau of Narcotics—passages the Justice Department obviously did not want in the edited version of the book. In urging me to now sign the agreement, Hundley assured me that these charges were under investigation and said: "Nick [Katzenbach] has to have something in writing to satisfy Narcotics and Alexander [Meryl Alexander, Director of the Bureau of Prisons]. Besides, it's good for you. You get the department's official stamp of approval." Hundley, who has since left the Justice Department, would later tell me, "I never thought they would use it to go back on their word."

Nor did I. Although I already had a copy of the manuscript, and Valachi was on record as desiring me to edit it, I did not simply want to get out any book; Valachi's story was so important, it presented so extensive and detailed a picture of organized crime in America, that I was determined that if it was to be done at all, it had to be done responsibly. While Valachi's text was an extraordinary effort for a man of his limited education, a great deal of clarification was needed that only he could supply, many areas had to be expanded far beyond what he had initially drafted, and considerable rewriting and organization of his material were required. All this needed his cooperation, which only the Justice Department was in a position to grant. I signed.

Then, on January 9, almost two weeks after the Associated Press story, the Italian-American newspaper *Il Progresso* printed a violent editorial denouncing the impending publication of "The Valachi Memoirs." This editorial was the first shot in a massive campaign to halt the book, and it set the pattern for all that followed. The editorial wondered out loud if the whole point of publication was to perpetuate the "kind of image of criminality associated with the many Italian names in Valachi's testimony" and then in the same breath noted, quite accurately, that it was an "image which millions of law-abiding Americans of Italian descent have consistently proven false through their outstanding achievements in the arts, in the sciences, in industry, in labor, in the professions, in government and in the religious orders."

The editorial, with an appropriate covering letter, was sent to every U.S. Senator and Congressman. "Professional" Italian-Americans around the country used it to keep the ball rolling. It was the same sort of outburst that greeted a widely acclaimed book, *The Italians*, by Luigi Barzini, himself a member of the Italian Parliament, because it did not present an idealized portrait of his countrymen, or the cries raised against Yale University for publishing a scholarly work which did not credit Christopher Columbus with discovering America. In Valachi's case the theme remained constant: his book would slur all Italian-Americans. P. Vincent Landi, representing the Order of the Sons of Italy in

America, declared that it was a "matter of civil rights" for Italian-Americans to have the proper image. Pennsylvania Supreme Court Justice Michael A. Musmanno added a new argument. Valachi, he said, would become a "millionaire" from the book, neglecting to explain how Valachi, doomed to life imprisonment and limited to $15 a month spending money, was going to enjoy all this imagined wealth.

There is no question that many of these fears were legitimately motivated; on the other hand, some protests were less than wholly disinterested. Among the loudest voices to be raised against publication of the manuscript belonged to those who subsequently formed the American-Italian Anti-Defamation Council, whose executive board turned out to be riddled with relatives and associates of Cosa Nostra mobsters. One board member, Daniel J. Motto, the head of a bakers union local, was recently convicted for playing a key role in a Cosa Nostra attempt to infiltrate the New York City government. It was Valachi, by now aware of these protests, who delivered the classic riposte. "What are they yelling about?" he asked me. "I'm not writing about Italians. I'm writing about mob guys."

Friends of mine in the Justice Department told me not to worry—that protests like this occurred every time it prosecuted a criminal who happened to have an Italian name. I first learned how serious the situation might be one Saturday afternoon in January while I was working with Valachi in the D.C. Jail. I received a telephone call from Rosenthal asking me to come to his office as soon as possible. When I got there, he said, "We've got problems," and showed me a batch of correspondence from various Senators and Congressmen to Attorney General Katzenbach, all of it referring to *Il Progresso*'s editorial. Some of it was intemperate; some was not. One letter, for example, from the highly respected Senator John Pastore of Rhode Island, simply requested background guidance in answering mail triggered by the editorial.

Rosenthal asked my help in drafting a reply to these letters. I did so. I emphasized the fact that the Valachi manuscript was a unique social document about a particular period of U.S. history; that, as would be made abundantly clear, it was not an ethnic story, but rather the experiences of one man; and that, as Gus Tyler, a leading social historian, has aptly pointed out in his introduction to one of the standard scholarly works in the field, *Organized Crime in America*: organized crime is not the "child of the Italians, or Irish, or Jews, or the Puerto Ricans—although it has drawn recruits from all these, and other, ethnic groups. Organized crime is a product and reflection of our national culture."

Rosenthal was confident that this reply would help work "things" out. By then, however, it was too late, and on February 1, 1966, a delegation of twelve Italian-Americans met in Washington with Katzenbach, Hundley, and Fred M. Vinson, Jr., Assistant Attorney in charge of the Criminal Division. In the group were four Congressmen, Peter W. Rodino, Jr., Joseph G. Minish, and Dominick V. Daniels, all from New Jersey, and Frank Annunzio of Illinois. Delegation members demanded that publication of the manuscript be stopped and threatened to take up the matter with the White House.

The late Senator Robert F. Kennedy, who, of course, previously had been Attorney General, offered me little encouragement when I discussed the situation with him. "They are going to go to Johnson," he said. "It's an easy way for him to get them on the hook. He'll tell Nick to kill the book, and Nick isn't going to argue with him."

While nobody in a position to do so is saying what transpired between President Johnson and Attorney General Katzenbach, I did learn that the protests were handled at the White House by Jack Valenti, a special assistant to the President. Shortly thereafter, the Justice Department reversed its entire stand on the Valachi book, although the key department people under Katzenbach who were involved in the affair were against any change. Vinson, until it became a matter of administration policy, was on record opposing it. Hundley, who had brought the department's drive against organized crime to an unparalleled degree of efficiency, told me that it was one of the reasons that led ultimately to his resignation. John Douglas, who, as Assistant Attorney General in charge of the Civil Division, would have to direct any court action to enjoin publication of the manuscript, sought vainly to find a compromise solution. Douglas, a man of recognized integrity, reflected the general embarrassment of the department when I tried to broach the subject with him considerably after the event. "Peter," he said, "I'm not even going to talk to you about this off the record."

Thus, on February 8, I received a telephone call from Jack Rosenthal. "I'm afraid," he said, "that the deal is going to be queered."

"What's happening?" I asked.

"We're just getting it from all over."

I told Rosenthal that it was inconceivable to me that the Justice Department, of all government departments, would knuckle under to such blatant political pressure.

"Well," he said, "it's not final or official, but things look terrible right now. I don't have much hope."

Once the decision to suppress the book had been made, I received an offer to reimburse me for the time and expenses I had put into the project. No dollar-and-cents figure was mentioned because I refused to discuss any financial settlement.

The late Senator Robert Kennedy, who was a friend of mine, then told me that Katzenbach, also a friend of his, had come to his office and asked him if he would "speak" to me. Kennedy declined to intervene, because he found it reprehensible that the Justice Department had gone back on its word, once it had authorized publication of the manuscript.

Convinced more than ever that this was a story that demanded telling, I retained attorneys to represent me. (Paul McArdle, whom the department had arranged to represent Valachi, found himself involved in much more than he had bargained for and would, with everyone's blessing, withdraw. As he said, he had not counted on "fighting the White House.")

On May 10, 1966, the Department of Justice asked the U.S. District Court in Washington to enjoin me from "disseminating or publishing" the manuscript, and a press release announcing this decision was hand-delivered to the four Congressmen who had been in the February 1 protest delegation.

It was the first time an Attorney General of the United States had initiated action to ban a book.

All at once, publication of a manuscript which the Justice Department had hailed as being of such benefit to law enforcement was suddenly "injurious" to this same law enforcement. If there ever was any doubt about the political pressure behind these moves, New York City Surrogate Court Judge S. Samuel DiFalco quickly dispelled it. DiFalco, as national chairman of the Italian-American League to Combat Defamation, dispatched a telegram to various Italian-American groups around the country which read: HAPPY TO INFORM YOU THAT AS A RESULT OF OUR PROTEST THE JUSTICE DEPARTMENT ANNOUNCED WITHDRAWAL OF PERMISSION FOR PUBLICATION OF VALACHI MEMOIRS.

I received, however, many letters expressing a different view. One was from Samuel J. Prete, chairman of the Italian-American Chamber of Commerce in Detroit, who noted:

"The very fact of the refusal of the Justice Department to permit the publication of Valachi's book is an indictment upon all Italian-Americans because it is an inference that all Italians are bad. I, as a good Italian-American, resent this."

On the West Coast the Italian-American newspaper *Corriere del Popolo* adopted the attitude considerably different from that of *Il Progresso*. "No one can hope to achieve through censorship anything meaningful," it pointed out. "The honest Italian-Americans have nothing to fear from revelations of Valachi. They have a social, ethical and moral standing in the United States that Valachi can never upset."

The general press, perhaps predictably, was almost unanimously sympathetic to the battle to free the book. As *Newsweek* wryly commented about the Justice Department's action, "It was hard to tell the good guys from the bad guys."

But now I had created the opportunity to express my views in open forum—in the court proceedings—for one way or another, despite the prospect of an extensive legal fight, I had resolved to publish Valachi's story. I lost the first round—contesting a preliminary injunction against publication obtained by the Justice Department—when the U.S. District Court in Washington ruled that the department would suffer "immediate and irreparable harm" if the injunction were not granted. The Court of Appeals upheld the preliminary injunction, emphasizing that it was "settled law" not to upset such an injunction by a lower court except in an instance of "clear error," and suggested that the case be brought to trial.

This was to be my next step, which would force the full story out into the open. I decided to go straight to the heart of the matter—the Memo of Understanding, which had been prepared by the Justice Department and which clearly indicated that it not only had proposed the publication of a book based on Valachi's manuscript, but had specifically emphasized the expectation that such publication would advance the ends of law enforcement.

To get on the record the facts behind this complete and abrupt reversal of policy, subpoenas were served for the purpose of obtaining pretrial depositions from Katzenbach, who by now had become Undersecretary of State, and other officials who had a pertinent connection with the case. The reaction was sharp and immediate. The Justice Department went to court to fight the taking of these depositions. During arguments, Judge Alexander Holtzoff suddenly addressed himself to the department attorneys:

It is one thing to have the right of editing or revising, but another thing to say to a person you can't publish at all. Isn't that rather shabby treatment? Well, perhaps I won't embarrass you by requesting an answer to that.

Here they allowed him [Peter Maas] to work on this manuscript. It may be it was improvident in the first instance. Then they say you can't publish it, we don't think you should publish it at all.

In return the Justice Department claimed in effect that I had not been treated all that unfairly, that I "had the opportunity to interview [Valachi] and to examine his life of crime as denoted by the manuscript," and that, while enjoined from publishing Valachi's manuscript, I was free to use all this material to publish a third-person book. This paved the way for *The Valachi Papers*.

The Valachi Papers is composed of my direct interviews with him, his written responses to many of my questions, interviews with people associated with Valachi, the several hundred pages of raw interrogation by the Bureau of Narcotics and the FBI to which I had access, and other official sources and documents.

Other than agents from the Bureau of Narcotics and the FBI, I am the only one who has interviewed Valachi. All during the protracted negotiations with the Justice Department, I continued to see him. My interviews often paralleled material he had written and discussed in public testimony and developed as well lines of questioning in areas he had left obscure. In many instances, simply to check the accuracy of his memory, I had him repeat his recollections on even such minor points as an experience in a Catholic reformatory, a police chase, an incident at a racetrack. He was like a tape recorder; it always came out the same.

In only one respect have I taken any liberties with what he had to say. Valachi's normal speaking vocabulary is rife with "nuttin'," "hadda," "gonna," and so on. In the interests of readability, these contractions were eliminated.

In all other respects it is the story they tried to suppress.

The
Valachi
Papers

1

At approximately 7:30 A.M. on June 22, 1962, at the U.S. Penitentiary in Atlanta, Georgia, prisoner number 82811, a convicted trafficker in heroin named Joseph Michael Valachi, seized a two-foot length of iron pipe lying on the ground near some construction work and, before anyone realized what he was up to, rushed at a fellow inmate from behind and in a matter of seconds beat him to a bloody, dying pulp.

Initially, however brutal the assault, it appeared to be nothing more than the kind of routine murder that periodically erupts under the stress of life inside prison walls. Certainly there was little to indicate anything exceptional about Valachi himself, at the time fifty-eight years old, a squat, swarthy, powerfully built man, 5 feet 6 inches tall, overweight at 184 pounds, with a thick crop of politician-gray hair, expressionless brown eyes and a guttural rasp of a voice, who, despite a career earnestly devoted to crime, was just about as obscure a hoodlum as one could hope to find.

But this seemingly senseless killing by apparently so unremarkable a hand would end with Valachi becoming the first person to unmask the Cosa Nostra, whose very existence had been a subject of fierce debate even in law enforcement circles. Almost overnight, as a result, Valachi's name became as familiar as that of a Capone, Luciano, Costello, or Genovese. Not only did he dominate newspapers, magazines, and television after some of the bizarre tidbits of what life was like with Vito, Joe Bananas, and Buster from Chicago were revealed, but he achieved true status when comics began cracking jokes about him.

There was, however, really very little to laugh at. Organized crime is America's biggest business. According to the best estimates of the Department of Justice, admittedly an educated guess, it grosses better than $40 billion a year. Even if such a staggering statistic was off by as much as half, it would still dwarf anything else in sight. Organized crime, of course, pays no taxes, but it does pay to corrupt countless public officials at all levels, and besides its lucrative illicit rackets, it has increasingly infiltrated and taken over legitimate businesses and labor unions—applying, naturally, its own ethical standards. While the Cosa Nostra does not embrace all organized crime, it is its dominant force, virtually a state within a state—a "second government" as Valachi puts it—painstakingly structured, an intricate web of criminal activity stretching across the nation, bound together in a mystic ritual that sounds like a satire on college fraternity initiations and at the same time caught up in a continual swirl of brutality, savage intrigue, kangaroo courts and sudden death.

Valachi lived in this world for more than thirty years without breaking its blood oath of allegiance—and silence. The circumstances that eventually caused him to do so began in Atlanta.

For weeks he had led a terror-filled existence. He was marked for death, and he knew it. Another prisoner, also a member of the Cosa Nostra, had accused him of "ratting" to the Federal Bureau of Narcotics. All at once Valachi found himself the target of the same sort of underworld execution that he had so often and so efficiently carried out in the past. While he had supplied narcotics agents with some fragmentary information about illegal drug traffic, in a not unusual bid for a lighter prison sentence, he ironically had never mentioned anything about the Cosa Nostra itself. Just what caused him to be fingered has yet to be entirely resolved. One theory is that the Bureau of Narcotics, convinced that he had a lot more to say about the movement of heroin across U.S. borders, deliberately spread the word to bring enough pressure on Valachi to break him down completely. A second theory, which Valachi, among others, subscribes to, is that his accuser, a codefendant in the same narcotics case, did it to divert suspicion from himself.

In June 1962, time was running out fast for Valachi. He had already survived three classic attempts to murder him in prison. One was to offer him poisoned food. Another was to corner him alone and defenseless in a shower room. Still a third was to goad him into a fight in the penitentiary yard, so that in the confusion of the rubbernecking crowd which would automatically gather around, he could be knifed.

Worse yet, he had no avenue of appeal. The Cosa Nostra is divided into major units, each of which is called a Family. Valachi belonged to one such Family in New York City ruled by Vito

Genovese, the most feared *capo*, or boss, in the Cosa Nostra. And it was Genovese, also in Atlanta serving a narcotics conviction of his own, who had decreed Valachi's death. At first everything seemed cozy between the two convicts, and Valachi could not believe that Genovese, who not only had invited him to become a cellmate and then arranged the move, but had been the best man at his wedding years before, would turn against him now. But all the warnings from friendly sources along the prison grapevine, as well as the hostile behavior toward him of other inmates currying Genovese's favor, were confirmed for Valachi in an eerie confrontation with Genovese. This is his account of what took place:

One night in our cell Vito starts saying to me, "You know, we take a barrel of apples, and in this barrel of apples there might be a bad apple. Well, this apple has to be removed, and if it ain't removed, it would hurt the rest of the apples."

I tried to interrupt him when he was saying this, but he waved at me to keep quiet. Finally I couldn't stand it no more. "If I done anything wrong," I said, "show it to me and bring me the pills—meaning poison—and I will take them in front of you."

He said, "Who said you done anything wrong?"

There wasn't anything I could say.

Then he said to me that we had known each other for a long time, and he wanted to give me a kiss for old time's sake. Okay, I said to myself, two can play the game. So I grabbed Vito and kissed him back.

After I did this, he asked me, "How many grandkids you got?"

I said, "Three. How many you got?" I think he said six. So I said, "It's good to know." In other words, if he's going to be concentrating on my grandkids, I'm letting him know I'll concentrate on his.

I went to my bed, and Ralph, who was in the next bed, mumbled, "The kiss of death." I pretended I didn't hear him and just laid on my bed. But who could sleep?

I didn't go for this "kiss of death" stuff. But I knew that just before a guy was going to be hit, the thing to do was to be very friendly with him, so as not to put him on guard. Now in the old days when you met another member, the habit was to kiss him. Charley Lucky* put a stop to this and changed it to a handshake. "After all," Charley said, "we would stick out kissing each other in restaurants and places like that."

On June 16 Valachi took a last desperate step to save himself and asked to be put in the "hole," prison lingo for solitary confinement. When the guard he had approached demanded to know why, he replied, "Someone's going to kill me, or get killed. Is that good enough reason for you?" Once in solitary, Valachi informed the prison's chief parole officer that he wanted to speak to George Gaffney, currently Deputy Director of the Bureau of Narcotics and former head of its New York office. Valachi's message to Gaffney—that he was "ready to talk"—was never relayed. In the inquiry that followed, the parole officer said that since Valachi refused to elaborate, he took no action on the ground that Gaffney would not make the trip to Atlanta without more information. Next Valachi wrote a letter to his wife. Through her he hoped to send word to another Cosa Nostra boss in New York, Thomas (Three-finger

---

*The late Salvatore Lucania, also known as Lucky Luciano, one of the chief architects of the modern Cosa Nostra.

Brown) Lucchese,* that the way in which he was being sum-
marily judged violated the Cosa Nostra code. His frantic letter
read:

I advice [sic] you that just as soon as you will receive this drop
everything and come and see me. Don't let money stand in your way.
It is most important. Don't waste one day. Understand. Then I will
never bother you any more. When you come make sure you get in.
Remember don't lose any time.

What Valachi planned was to fence with Gaffney at least long
enough to allow Lucchese, who had been friendly to him over the
years, to intervene in his behalf. But the letter never left Atlanta.
Instead, according to Associate Warden M. J. Elliot, it was returned
to Valachi for rewriting on the chance that this might give prison
officials some insight into why he had requested solitary confinement. Valachi would not do it. By now he had concluded that he no
longer could trust the prison administration. And from his standpoint the idea was perfectly reasonable; to him the power and
influence of a Vito Genovese were limitless.

Ordered out of solitary because of his continued silence,
Valachi decided to act on his own. He would die, perhaps, but he
would take with him as many of those in the cabal to kill him as he
could. High on his list was a veteran Cosa Nostra enforcer and
Genovese crony, Joseph (Joe Beck) DiPalermo. At no time, however, did he consider striking directly at Genovese. For Valachi,

---

*Such aliases are common in the Cosa Nostra. Lucchese, now dead, acquired
his in 1915 when he lost his right index finger in an accident. At the time there
was a well-known major league pitcher called "Three-finger" Brown because
of a missing digit.

fantasizing like an angry child, "Vito must live" to stand exposed before the rest of the Cosa Nostra as someone who sentenced his men to death without any sort of a hearing and then dismissed it by calling them "rats."

Everything came to a head on the morning of June 22. Half out of his mind under this grinding pressure, and having eaten practically nothing for days for fear of being poisoned, Valachi finally exploded against a man he thought was DiPalermo:

I was out in the yard down by the baseball diamond. All of a sudden I saw three guys behind the grandstand looking at me. They were about fifty yards away. Then they started towards me. I had my back against the wall. There was some construction work going on, and I saw a piece of pipe lying on the ground. Just as I picked it up, figuring that if I'm going to go, they're all going to go, a guy walked by and said, "Hello, Joe." I looked up as he had passed me. He looked just like Joe Beck, so I said to myself, I might as well take him, too. I took the pipe, and I let him have it over the head. He fell. Then I ran after those three guys. One of them had a knife. I went about ten yards, when they turned and started to run away, so I ran back to the guy on the ground. With all the blood, who could tell who the hell he was now? I gave him two more shots with the pipe.

The three of them from the grandstand started running back towards me. They were about twenty yards away when a guard ran up and told me he wanted the pipe. I said I wouldn't give it to him.* He kept after me. I said, "Leave me alone or I'll use the pipe on you." He said, "There's a guy

---

*Guards within physical reach of prisoners do not carry weapons lest they be overpowered and the weapons seized. Guards on the wall, however, are armed.

dying." I said, "Good. Let him die," thinking all the time it was Joe Beck.

Now there were about twenty inmates around us. The guard said, "Let's go to the associate warden's office." I said, "Okay, but I'm keeping the pipe."

It was in the associate warden's office that I found out I got the wrong guy. I was there about fifteen minutes when he went out of his office for a second and came back in and threw a picture at me. "Do you know him?" he said. I said, "No," and he said, "Well, that's the man you just hit."

I didn't know what to think. I was in a fog.

The object of Valachi's raging attack was named John Joseph Saupp, in Atlanta for mail robbery and forgery, a man with no organized crime connections, whom Valachi did not even know, but a man who bore a remarkable, and fatal, physical resemblance to his intended victim, DiPalermo. This, in the opinion of a special agent of the FBI who would later spend more time with Valachi than anyone else, was the turning point. "Valachi," he says, "has no real remorse for anything he has done in his life, except this. Nothing crushed him more than the fact that he got the wrong man. It really plagues him. Getting a guy who was going to get *him* was the one satisfaction he was willing to settle for. If he had been successful, he probably never would have talked."

Saupp, despite multiple skull fractures, lingered on without regaining consciousness for almost forty-eight hours before he died. Even though Valachi was now facing a murder charge, he still insisted that "I just went crazy" to prison officials who questioned him. As one of them unprophetically noted at the time, "I get the feeling that Valachi . . . will never come out with a full account of the whole story."

prosecuted for dealing in narcotics, knew what had happened. Meanwhile, the local U.S. Attorney in Atlanta, who had jurisdiction over the case, was preparing to ask the death penalty for Valachi because of the "brutality and senselessness" of the killing. At this critical juncture Valachi himself finally managed to get word of his plight to Robert Morgenthau, U.S. Attorney for the Southern District of New York, via a go-between whose name, for obvious reasons, is still being kept under wraps by the Justice Department. A phone call to Morgenthau on July 13 briefly sketched the situation and advised him that Valachi "now wants to cooperate with the federal government." Morgenthau, after a hurried conference with the New York office of the Bureau of Narcotics, immediately contacted his counterpart in Atlanta and informed him of Valachi's potential value. The upshot was that on the morning of July 17, Valachi, represented by two court-appointed lawyers, was permitted to plead guilty to a lesser charge of murder in the second degree and received a life sentence. That same day he was taken in tow by narcotics agent Frank Selvagi and flown to the Westchester County Jail, a few miles north of New York City. There he was given the cover of "Joseph DeMarco" and installed apart from other prisoners in the jail's hospital wing.

But the hope that he would promptly spill everything he knew soon vanished. At the time he murdered the wrong man in prison, Valachi was serving concurrent terms of fifteen and twenty years for peddling dope. While he freely admits to a number of specific crimes, including complicity in several previously unsolved gangland killings, he still maintains that on the second of these convictions, which led to his being labeled an informer, he was framed. So, once out of the reach of Vito Genovese, he took out all his hostility on agent Selvagi. "You were the cause of me getting into trouble," he jeered. "Where were you when I needed you?"

Valachi nonetheless was careful to hint at just enough to keep Selvagi interested. And the Bureau of Narcotics, angling at best for inside information on heroin traffic, got a peek at considerably more than it bargained for. Gradually the shadowy outlines of a national crime cartel involving a vast array of rackets began to emerge out of the give-and-take between Selvagi and Valachi. Toward the end of August, as a result, Henry L. Giordano, the Commissioner of the Bureau of Narcotics, called a key Justice Department figure, William Hundley, who headed a special section set up by Attorney General Kennedy to pull together the previously uncoordinated efforts of various government investigative agencies against organized crime and racketeering. "We're talking to a guy," Giordano told Hundley. "It could be important. I'll send you copies of the reports we're receiving."

Almost at once Hundley found himself in the middle of some bitter bureaucratic infighting. The Federal Bureau of Investigation, which had gotten wind of what was going on, asked him for a look at the reports. Afterwards, according to Hundley, FBI interest in Valachi became "overwhelming." The FBI formally requested access to Valachi on the ground that the information he had to offer "transcended" traffic in narcotics. Informally, Hundley was told, "We have to have him." Such pressure was startling. "It puzzled me," Hundley recalls. "Here suddenly was the most prestigious law enforcement agency in the world all worked up over one man. They usually don't get that excited."

But Valachi meant a great deal to the FBI for a very special reason. Prior to the advent of Kennedy as Attorney General, it had been paying little attention to organized crime. In 1959, for example, only four agents in its New York office were assigned to this area, and their work was primarily in-office "bookkeeping" chores collating such routine information as the whereabouts of known racketeers. On the other hand, upwards of 400 agents in the same office were occupied in foiling domestic Communists. Although FBI Director J. Edgar Hoover nominally takes orders from the Attorney General, he had operated under a succession of them as if they never existed. Kennedy was able to change much of this, at least during his tenure. Not only was he knowledgeble and concerned about organized crime and determined to crimp its mushrooming growth, but he also had a brother in the White House. Thus by 1962, again using New York as an example, about 150 agents, the bulk of them drawn from security work, were specializing in organized crime, assigned to specific cases, ferreting out leads, actively engaged in surveillance, etc. Still, embarrassingly caught off guard by Kennedy's initial demand for underworld intelligence data, the FBI high command had been forced to resort to widespread wiretapping and bugging to provide information which it euphemistically ascribed to sources like "confidential informant T-3, known to be reliable in the past." Now all at once here was Valachi, the first warm body to come forward whose statements jibed with this electronic eavesdropping and apparently a hot prospect to fill in the gaps.

With Kennedy's backing, however, Hundley decided to give the Bureau of Narcotics more time to develop Valachi on its own. But two weeks later, when a progress report from Selvagi proved disappointing because of Valachi's continuing hostility toward the Bureau of Narcotics, the FBI request was granted, and a crack special agent from its New York office, James P. Flynn, entered the picture ostensibly to delve further into the circumstances of the Atlanta murder. For a time Flynn and Selvagi jointly questioned

Valachi, who was in his glory playing one agent off against the other. He had yet to mention the Cosa Nostra by name, and there have since been charges that he fabricated it. The fact is that the words "Cosa Nostra" had been cropping up on wiretaps, and at least a year before Valachi came along the FBI had reason to believe that it represented what was commonly called the Mafia.

Valachi dramatically confirmed this. Alone with him for a brief period on the afternoon of September 8, Flynn suddenly said, "Joe, let's stop fooling around. You know I'm here because the Attorney General wants this information. I want to talk about the organization by name, rank, and serial number. What's the name? Is it Mafia?"

"No," Valachi said. "It's not Mafia. That's the expression the outside uses."

"Is it of Italian origin?"

"What do you mean?" Valachi parried.

"We know a lot more than you think," Flynn said. "Now I'll give you the first part. You give me the rest. It's Cosa—"

Valachi went pale. For almost a minute he said nothing. Then he rasped back hoarsely, "Cosa Nostra! So you know about it."*

According to Hundley, Flynn became the indispensable figure in getting Valachi to talk. "Without him," Hundley says, "we could have blown the whole thing. Flynn is an unusual agent with

---

*It can be argued that Cosa Nostra is a generic, rather than a proper, name. Although the literal translation is "Our Thing," Valachi, when referring to it in English, would do so in a lower-case sense—e.g., "this thing of ours." There is also evidence that other terms are used in the United States. For instance, while the structure of a Cosa Nostra Family in Buffalo is identical to one of its counterparts in New York City, it is known locally as "the arm." It is really an academic question since, whatever the term, it adds up to the same thing.

great imagination and initiative. He had a tremendous knack of winning Valachi's confidence. He knew exactly when to be tough and when to baby him along. If Valachi was sick, for instance, he was the one who would bring him his medicine. If Valachi went into one of his depressions, he would always come up with the right thing to snap him out of it. Every so often he would bring some of the delicacies—cheeses and spiced sausage—that Valachi liked. These may sound like small things, but they made all the difference. Flynn practically lived with him for eight months, and Valachi wound up thinking he was the only friend he had—and quite frankly, Valachi was right."

A successful interrogation is a complex art. Some measure of the problem of dealing with Valachi is contained in a probation officer's report written in 1960 while he awaited sentencing after his last narcotics conviction. "There is little to be said in his favor," the report concluded, "since he has failed to demonstrate any real semblance of moral conscience and social conformity. He has never been quite in tune with the society in which he lives, and at this late date there is little reason to indicate that he ever will."

For Flynn the key was to isolate the motivations that finally led Valachi to talk and to play on them constantly. "Revenge was a large part of it," he later noted, "but it was also a cold, calculated move for survival. Don't think for a moment that this was a repentant sinner. He was a killer capable of extreme violence. He was devious, rebellious against all constituted authority, and he lived in a world of fear and suspicion. Fear especially marked him. Fear of what he was doing and at the same time fear that nobody would believe him."

By the end of September Valachi was placed entirely in the FBI's hands. And until the following January Flynn, subsequently joined by another FBI agent, questioned Valachi at the Westchester County Jail on an average of four days a week. A typical session lasted about three hours, after which Valachi tended to become increasingly jumpy and difficult to manage. During this period it was soon established that Valachi's remarkable memory worked best when he was allowed to tell his story in stretches without interruption even though any one sentence might feature a half dozen unidentified "hes," "hims" and "thems" which had to be tracked down in later interviews.

After the initial breakthrough, Valachi sketched the broad organizational structure of the Cosa Nostra with comparatively little prodding. And for the first time the Justice Department got a picture of its enormous scope. He revealed that the Cosa Nostra is divided into Family units, each supreme in its own area. These areas include Boston, Buffalo, Chicago, Cleveland, Detroit, Kansas City, Los Angeles, Newark, New Orleans, New York City, Philadelphia, Pittsburgh, and San Francisco. In Valachi's words, resort centers such as Miami and Las Vegas (or pre-Castro Havana) are "loose" or "open." This means that any Family, regardless of its base, can maintain members and conduct operations there.

The ruler of a Family is known as the *capo*, or "boss." Next in command is the *subcapo*, or "underboss." Then came a number of *caporegime*, or "lieutenants." Each lieutenant is in charge of a *regime*, or "crew." A crew in turn is composed of "soldiers," whose actual status depends a good deal on their individual experience, contacts, and ability. Some soldiers, for instance, work directly for their lieutenants. Others run their own rackets. From boss to soldier, however, the members of the Cosa Nostra are united by one great common bond: they must be Italian. But by no means does the Cosa Nostra ethnically limit its operations. It is a closed society within a large framework, constantly involved with a whole spectrum of "outsiders"—Jewish, Negro, Irish, French, Puerto Rican, English, and so on down the line.

About a third of the Cosa Nostra membership is in New York City, which is unique in having five Families. Valachi named the boss of each. While police had long recognized them as major racketeers, their exact position in the underworld was uncertain. All pretended to be respectable businessmen, and all were so swathed in protective layers of underlings that they hadn't seen the inside of a jail cell for years. Indeed, so exceptional was the conviction of Vito Genovese that the Cosa Nostra universally believes that he was framed. Genovese, who purports to be nothing more than your friendly scrap dealer, was asked once if he had any moral scruples about killing a man. He gave the stock underworld reply: "I respectfully decline to answer on the ground that my answer may tend to incriminate me."

A second Cosa Nostra boss in New York City named by Valachi is Joseph (Joe Bananas) Bonanno. In an effort to achieve legitimate status, Bonanno took up residence in Arizona, where he posed as a successful dabbler in real estate until he was forced to return to quell a mutiny in his Family. Still a third, Carlo Gambino, blandly describes himself as a labor-management "consultant."

Two of the bosses Valachi identified have since died. One was Joseph Profaci, ostensibly an olive oil importer, who ran his Family in Brooklyn for more than thirty years and whose flower-wreathed funeral was in the best underworld tradition. The other, Thomas Lucchese, conceded only to being a prosperous dress manufacturer. Lucchese's last recorded conviction was for grand larceny in 1923. A few years later the wife and mother of a murder victim identified him as the killer, but before he was brought to trial, they changed their minds. One can sympathize with them; when Vito Genovese faced a murder charge of his own, the chief witness against him, despite being held in protective custody, was poisoned.

On a national scale, the Cosa Nostra has no Mr. Big, although Genovese was bidding for the title until he was sent to prison. Instead, it has been governed in recent years by a *Commissione* of from nine to twelve bosses across the country. The *Commissione* has one main function: to keep the Cosa Nostra a going concern. It is the final arbiter on disputes between Families. And when a boss dies or is otherwise removed, it must confirm the man who takes his place. Thus, Valachi revealed, one of the reasons for the celebrated 1957 Cosa Nostra conclave near the little town of Apalachin, New York, was to bless Carlo Gambino as the successor to Albert Anastasia—familiar to tabloid readers as the Lord High Executioner of Murder Incorporated—after his bloody demise in a Manhattan barbershop.

At the Westchester County Jail, having gone this far, Valachi with typical ambivalence began trying to learn how much the FBI really knew about the inner workings of the Cosa Nostra. Flynn spent endless hours in this sort of sparring. Each time it happened, he would patiently say to Valachi, "You must be our teacher. It doesn't matter what we know. You must tell us."

Flynn also repeatedly warned Valachi that he must tell the truth and that if caught in even one lie, it would discredit his entire story. Sometimes Valachi sulked at such a suggestion. Other times he would blow up. Once he shouted back angrily, "The truth! The truth! You're always yelling about the truth. What the hell do you think I'm telling you?"

These warnings nonetheless had the desired effect, and Valachi followed a consistent pattern. If he described a situation that he knew about personally, he would make a flat declaration. If the information was secondhand, he would preface it by saying, "One of the boys told me." If Valachi believed it himself, he would refer to the source as "solid."

The tension Valachi was under was enormous, and his moods took violent swings up and down. Scarcely a day went by without some reference to Vito Genovese and the death sentence he had decreed. When Valachi was depressed, he would complain, "That bastard! Why did he do it?" When he was up, he would laugh bitterly and say, "I was as smart as he was. I was waiting to see his moves while I was making my own." On one occasion he told Flynn, "I'm a dead man already. I know it. But the more I live, the more shame it is for Vito."

In his blackest moments, pacing ceaselessly back and forth, chain-smoking cigarettes, he dwelt on the futility of trying to break the power of the Cosa Nostra. "What good is it what I'm telling you?" he would cry. "Nobody will listen. Nobody will believe. You know what I mean? This Cosa Nostra, it's like a second government. It's too big."

To counter this, he was constantly reminded that in Kennedy and Hoover he had some heavyweight help on his side. There were also elements of his own character that could be capitalized on. One was his habit of blaming his troubles on everyone else. "Joe," Flynn recalls, "thought everybody was responsible for Joe, except Joe. He only took up crime, for instance, because he never had a chance as a kid. It was the prison administration's fault that he had to kill a man he didn't even know. He doesn't even consider himself a traitor to the Cosa Nostra; in his mind Vito Genovese was the real traitor."

Another characteristic that proved immensely valuable was Valachi's frustrated ambitions within the crime syndicate. "Why," he would often complain, "should some guy who put in less time be ahead of me?" Still a third was his need for ego satisfaction. "You say you want to destroy the bosses," he was told. "This is the way you can help do it."

Thus alternately coaxed and prodded, Valachi continued to describe his life in the mysterious and violent underworld of the Cosa Nostra. As he did, a remarkable physical transformation began to occur in Valachi. "In the beginning," one of his custodians remembers, "he was fat, maybe forty pounds too heavy, and he looked like a bum. He was a guy without pride. All he wanted to do was exist. We had to order him to change his clothes, and he went for days without shaving. All of a sudden he started going on a diet. Then he asked if he could have some exercise equipment, and he would work out at least an hour every day. It was like he had gone into training for a fight."

His spiritual catharsis, however, never quite kept pace with his physical rebirth. A Roman Catholic, he was asked several times if he wanted to see a priest. He always refused. "I got no time for that," he would say. On the other hand, he displayed one unexpected bit of gallantry in refusing to name any of the women in his life. When the subject came up, he said, "Let's just say that I never abused a girl. So don't ask no questions. What's my business is mine."

The only deal federal officials ever made with Valachi was an agreement "to leave his family out of it." Valachi is married and has one son. His wife, Mildred, is the daughter of a Cosa Nostra boss named Gaetano Reina who was gunned down in 1930. His son is in construction work in New York. Valachi, who had various legitimate fronts, which included three restaurants, a dress factory, and a jukebox operation, says his son knew nothing of his criminal activities. "The kid," as he once put it, "is a square. I didn't want him to know nothing about me. I didn't want him in the life."

During his interrogation nothing fed Valachi's ego more than to play the prima donna, although he was always careful not to overdo it. The tip-off came at the beginning of an interview session when he would grandly announce, "I don't want to talk about anything today. I want to relax." A rabid Yankee fan, he would inevitably start discussing baseball. Or he would reminisce. Once he described his first crime to Flynn. As he told it, his father could not meet the rent. So Valachi, who was nine years old, and his brother stole a crate of soap cakes from a neighborhood grocery and sold them door to door at half price. With his amazing memory, he even recalled the brand name, Fairy Soap, and seemed disappointed that Flynn had never heard of it.

He was supplied with his favorite reading matter, the *New York Daily News,* every morning. The first thing he would do was turn to the racing results. As the onetime owner of four thoroughbreds, he spent his spare time handicapping the horses, and Flynn always knew he was in for an especially productive day whenever Valachi told him that he had picked "some winners." Next he would scan the obituary page. It occasionally gave him a chance to display a morbid sense of humor. One morning he found the death notice of a Cosa Nostra member whom he had collaborated with on a murder assignment. In between chuckles Valachi recounted how his late friend's trigger finger had "frozen" after he had lined up his intended victim.

For all this, he never lost his instinctive guile. In his Cosa Nostra days, when he was handed a "contract" to kill, Valachi went to great lengths to make sure it had been formally authorized and not one, as was often the case, motivated by personal greed or revenge which could land him in hot water between rival bosses. At the Westchester County Jail he applied the same technique. The place, he suddenly began complaining to Flynn, was giving him the "creeps." For one thing, there was a prisoner there who had known him in the "old days." For another, "everybody" knew the jail was regularly used by the Bureau of Narcotics to question informers, and it would be just a matter of time before Cosa Nostra tracked him down. Then, applying the clincher, he protested, "If the boys knew I was in a dump like this, they'd laugh me out of town."

No one doubted that Valachi's real motive was to test the interest and power of Attorney General Kennedy, but it dovetailed nicely with plans to take him out of Westchester anyway. After weeks of looking around the country, the FBI had finally made arrangements with the Army to lodge him temporarily at Fort Monmouth, New Jersey, ideal because of its tight security as a communications center and also because of its proximity to Flynn's residence. For both Valachi and the FBI, Monmouth accomplished another cherished objective: it placed him completely beyond the reach of the Bureau of Narcotics.

(Still, Valachi's fear about Westchester were not so far-fetched. The FBI has since established that the Cosa Nostra had traced him to the New York area. His former associates, however, believed that he was being hidden in a Manhattan hotel and concentrated their search in that direction.)

Once the Justice Department's William Hundley authorized the move, Valachi was brought by federal marshals to Fort Monmouth in January 1963. He went by car, wearing civilian clothing obtained from his home by an FBI agent. "He looked like a hood on the late late show," one of his escorts says, "wide-brim hat, collar tabs halfway down his chest, and a cashmere polo coat with the biggest lapels I ever saw. He made quite a sight." At Monmouth Valachi was put in the stockade under the immediate protection of hand-picked guards from the Federal Correctional Institution at Lewisburg, Pennsylvania. The head of this detachment, told only that he would have a prisoner in his charge requiring maximum security, chose the Army's version of it—a tiny enlisted man's cell completely surrounded by bars. Valachi took one look and blew up. "What the hell is this," he raged, "some kind of fucking cage?" He was finally mollified the next day after agent Flynn arrived on the scene and saw to it that he was shifted to more private quarters in the stockade reserved for officers.

In a way the incident served to reinforce Valachi's confidence in Flynn and the FBI, and from then on his interrogation proceeded without a serious hitch. Valachi had been scheduled to stay at Fort Monmouth for three weeks. But it required the rest of the winter to cover the gaps in his story, to recheck key episodes, to gather, as Flynn put it, the "fine type" of information necessary to substantiate what Valachi had to say, and to pin down the Cosa Nostra membership—who precisely was in it and, just as important, who was not. This phase of interviewing was handled with extreme care. If, for instance, Valachi was describing a gangland killing, he had to supply detail upon detail for later comparison with the data contained in FBI or police files. As for all the members of the Cosa Nostra he personally singled out, Valachi first had to identify them from photographs either by name or by a known alias.

By now he also had expressed a willingness to testify in public. Three key points caused Attorney General Kennedy and his staff to favor the idea. One was the fact that Valachi had never said anything that conflicted with fragmentary information already known to the Justice Department. Another was the possibility that his testimony might create support for dormant anticrime legislation and could perhaps pave the way for legalized wiretaps under stringent federal court control. The third was the belief that Congress and the country ought to hear, as only an insider could tell it, what had been privy to only a handful of law enforcement officers.

While this was being considered, however, the question of whether or not to keep Valachi's revelations secret became academic. The Bureau of Narcotics, having lost control over his interrogation, feared that it would never get any credit for breaking him down. As a result, the bureau began to leak some details of Valachi's story, emphasizing its own role in the affair, to a friendly Washington reporter.

That settled the matter. And on September 9, after considerable preparation, Valachi was flown by helicopter to the District of Columbia Jail in Washington to face an investigating subcommittee chaired by Senator John McClellan of Arkansas. Valachi was greatly impressed by the flight and the security steps taken to assure his safety; since word of his presence at Fort Monmouth had gotten out by then, both he and his guards were disguised as Army MP's when they left the stockade. He was less impressed when Senator McClellan visited him privately at the D.C. Jail, just before the hearings began. According to Valachi, he requested he please skip any mention of Hot Springs, in McClellan's home state, and the Senate testimony contains no reference to that then-notorious city.

Valachi's televised appearances before the subcommittee were a disaster. But Valachi was the least to blame. Flustered by the circuslike atmosphere, he was out of his depth. While he does have nearly total recall, his mind is not the kind that jumps quickly from one subject to another. Yet the Senators almost to a man constantly broke in to ask wildly disparate questions. Even worse, although they all had been briefed on the specific areas of Valachi's knowledge, the minute the lights went on, he was bombarded with queries designed to appeal to the voters back home.

Perhaps the high point of the nonsense occurred when Nebraska's Senator Carl Curtis asked Valachi about the state of organized crime in Omaha. After a moment's reflection Valachi carefully cupped his hand over his mouth, turned to a Justice Department official sitting next to him, and whispered something. Those viewing the scene could be forgiven for supposing that Senator Curtis had hit on a matter of some import which Valachi wanted to check out before answering.

He was in fact asking, "Where the hell is Omaha?"

2

I first met Joseph Valachi on January 6, 1966, at the D.C. Jail, where he was still being kept on the top floor in what is nominally the death house, although no execution had taken place there for years. To get to Valachi, once inside the jail proper, I had to pass muster at an electronically controlled gate. A phone call then was made to one of the U.S. Bureau of Prison guards specifically assigned to be with him around the clock. Next I was escorted down a corridor to a heavy steel door with a massive key lock and taken up three flights of stairs to still another steel door, featuring a glass peephole through which I was again observed before being allowed into Valachi's quarters. These security precautions were not designed to keep him from escaping. They were there to keep him alive against the $100,000 price tag that has been placed on his head by the Cosa Nostra.

My initial reaction on seeing him in person was surprise that such a bandy-legged little man could cause so much commotion. Instead of the usual prison outfit, he was wearing a gray sweat suit, one of the two that made up his wardrobe, white athletic socks and gym shoes. Down to 145 pounds from his Atlanta days, he continued working out religiously with an isometric exerciser. His belly was flat and hard, and he had bulging shoulder muscles. Although he was then sixty-three, he could do thirty push-ups with little effort. His only vice was smoking sixty-odd cigarettes a day. "I know I shouldn't do it," he told me, "but what else can I do in here?" Still, it bothered him and he had shifted from Camels to Salems on the theory that they were less harmful to his health. "What do you think?" he asked. "Do you agree?"

I saw Valachi in the D.C. Jail twenty-two times from January 6 through March 19 for four to five hours at a stretch. Following each session, I also left him questions which he would then answer in longhand on the legal-size note pads he favored. His writing style was simple, declarative, and almost devoid of punctuation. He was apologetic about what he felt were literary shortcomings in these responses. "You got to remember," he told me, "I only made it through the seventh grade. But I'll say I'm a 65 percent better writer than when I started all this."

These interviews all went quite smoothly with one exception. He became extremely upset when I questioned him about the use of strong-arm methods in his loan-shark operations. This had been prompted by an FBI report that he often carried a baseball bat with him while making the rounds to delinquent debtors. But except for "maybe two or three times," he was outraged at the suggestion that he ever used force to collect loans. "What did I want to do that for?" he snapped. "I only dealt with people I knew would pay—bookmakers, policy men, guys like that. I stayed away from businessmen. What good was it beating up somebody? The idea was to keep the money moving."

He was as adamant about his role in underworld slayings. With the rebellious personality traits that would have made it difficult for Valachi to fit smoothly into any organization, much less the Cosa Nostra, the Justice Department is convinced that he lasted as long as he did because he performed a vital function as an expert killer in, according to its count, at least thirty-three murders. While Valachi doesn't deny being involved in many such "contracts,"* he contended that particularly in his younger days his specialty was to drive the getaway car and that with few exceptions he was never actually present when a victim was shot or otherwise dispatched. Later in his career, he insisted, his job was to supervise all the details of a hit, so that even in those instances when he acknowledged being physically on the scene, someone else always did the deed. No matter how I pressed him, he would not budge on this point. "Why should I lie about it?" he argued with disarming logic, as well as an eye perhaps cocked on his public image. "What do I gain? Legally I'm just as guilty."

Valachi had free run, such as it was, of the death house at the D.C. Jail. A corridor led past a guard's cubicle to a row of three cells for condemned prisoners and a shower room, where he had set up the one tangible reward accorded him for talking—a hot plate on which he could turn out a meal of his own. His foodstuffs, stored in a small refrigerator, were furnished by a couple of his special guards, an occasional visitor from the Organized Crime and Racketeering

---

*A "contract" is a common underworld term for an assignment to murder a designated victim. The murder itself is often called a "hit."

Section, or an FBI agent who wanted more information about a particular racketeer. Valachi was such a good cook that his guards asked him to write his recipes for their wives. He entitled his most popular creation "Joe's Special Recipe for Spaghetti Sauce and Meatballs." All his recipes were full of helpful hints. He noted, on the subject of chopped meat, "If you buy a pound of beef and grind it yourself, that's better yet, as you will know it is fresh." He was also careful to advise the novice cook that in making meatballs, "As you roll the meat on the palms of your hands, you keep putting a little olive oil on your palms, to prevent the meat from sticking." Like any good teacher, he went out of his way not to stifle the creative impulses of his pupils. "Make the meatballs," he exhorted, "any size you want."

Valachi spent most of his time in a narrow room at the other end of the death house corridor, and it was here that I would interview him. The room had an army cot, two chairs, a writing table, a radio, and a television set. Valachi's feet suffer in cold weather, and he had crudely Scotch-taped the louvers of the door opening on the corridor in an effort to ward off drafts. He also had affixed a large piece of muslin to the wall over his cot with Scotch tape. Curious about this, I peeled part of it back one day when I was alone for a moment. Valachi had put it up for understandable reasons. The muslin masked a rectangular viewing glass; on the other side was an electric chair. The room had been formerly used to witness executions. The only time he ever directly acknowledged the chair's existence was when I was photographing him and asked him to pose alongside it. "No," he said emphatically, "not me. It's bad luck."

Valachi's normal routine day after day at the D.C. Jail was exercise in the morning, a nap after lunch followed by small talk or a card game with one of the guards who rotated on duty in the death house with him, and writing letters or watching television at night. Federal prisoners on good behavior are given television privileges; Valachi had his own set simply because of his isolated confinement, and on it he had seen his appearances before the McClellan subcommittee. The show that offended him the most was *The Untouchables*. "It's all wrong," he told me. "How can they put on something like that?"

Valachi always maintained a certain pride. While I was an obvious source for some of the delicacies he fancied, he never asked me to bring him anything until I suggested it. Finally he agreed that I, as a New Yorker, might be able to get him some items not readily available in Washington. One was *capicole*, a spiced Neopolitan ham. Another was a cream pastry called *cannoli*. When he mentioned the *cannoli*, I told him that only a few nights before I had been to a first-rate Italian pastry shop on Manhattan's Grand Street, but I couldn't recall its name. As soon as he heard Grand Street, Valachi grinned and said, "That's Ferrara's. You're right. It's the best. Ah, you can't beat New York."

The same pride asserted itself once when Valachi flared at me. Throughout our talks the word "respect" had cropped up regularly. It symbolized the whole code of courtesy with all its subtle variations that exists in the Cosa Nostra regulating the behavior of Family to Family or member to member. One afternoon I arrived at the D.C. Jail to work with Valachi without having given him advance warning. His displeasure was quickly evident. When I asked him what was wrong, he said, "I didn't know you were coming. I would have shaved." I told him that it didn't make any difference to me. "Well," he said, "I want to show some respect."

Like almost everyone who spent any time with Valachi—

guards, Justice Department lawyers, FBI agents—I grew quite fond of him. Indeed, after always greeting me with solicitous inquiries about my health, the well-being of my family, and my comfort during the flight down from New York, Valachi would sit opposite me, hands folded over his stomach, a pair of reading spectacles perched haphazardly on the end of his nose, and take on a positively benign appearance. All that was really needed to complete the picture was a circle of grandchildren around him waiting to hear an amusing story, and I never ceased to be startled at hearing myself say something like, "Joe, the last time you were telling me how Steve Franse was strangled in your kitchen. Could we go over that again?"

Rather than return him to prison, plans were in the works to transfer Valachi out of the country, most probably to the Turk Islands in the western Pacific, once held by the Japanese and now under U.S. control. This satisfied the requirement that he be both confined and protected and yet would reward his cooperation. The Justice Department's Organized Crime and Racketeering Section, anxious for more Cosa Nostra informers, was especially high on the idea since one of the critical elements in the intelligence Valachi supplied was a list of those among his former colleagues who might be persuaded to talk.

Everything went by the boards, of course, when the White House, bowing to Italian-American pressure groups, determined to suppress publication of Valachi's story. And on March 22 he was suddenly yanked out of the relative comfort of the D.C. Jail and put back into the federal prison system at Milan, Michigan, forty miles southwest of Detroit. He was placed in a cell just big enough to pace eighteen feet in each direction. Its concrete walls were painted light blue. Three barred windows afforded a view of some of the Milan Prison's dingy red-brick buildings. A partition extending out about eight feet divided the cell in half. On one side there was an open toilet, a washbasin, and a shower stall. On the other side was an iron cot, a table with a radio on it, a television set, two chairs, a small wooden cabinet, and, draped over the radiator, his exerciser.

Here he was to have spent the rest of his life in solitary confinement.* When a member of the Organized Crime and Racketeering Section first heard the news, he slammed his fist in furious frustration against the wall. "That is a great way to develop informers," he said. "I can just see myself now telling some guy we're trying to get to talk that he doesn't have to worry—that we'll take care of him."

In the unlikely event that Valachi is ever considered for parole, the earliest it could happen is in 1980, when he would be seventy-six. As a federal prisoner, he is limited to $15 a month spending money. Since, because of his isolation, he cannot earn this money as other convicts do in the prison workshops, he has been dependent on contributions from various people who started corresponding with him after his defection from the Cosa Nostra; they are, in the main, women who hail him for his "courage" or his decision "to make up for his past." The only other financial worry that has concerned him is his fear of receiving a pauper's burial.

One factor at the Milan prison remained the same as it had been at the D.C. Jail. He was surrounded by extraordinary security measures to guard his life. An almost endless series of locked doors cut him off not only from the outside world, but the rest of the

---

*Two Michigan winters, however, proved too much for Valachi's health, and in July 1968, he was transferred to the Federal Correctional Institution at La Tuna, Texas, near the New Mexico border.

prison. His food tray, instead of being specially prepared for him was selected at random from the mess hall. Besides the normal bars, his windows were also covered with heavy wire mesh.

There was, however, another threat to his life that nobody had counted on—from Valachi himself. And early on the morning of April 11, 1966, he ripped out the electric cord of his radio, stepped into his shower stall ostensibly to bathe, knotted one end of the cord to the shower head, and tried to hang himself.

That he failed was pure chance. A badly finished metal disk on the shower head had an edge just sharp enough at the right spot to slice through the cord under the dead weight of Valachi's dangling body. Then a guard on duty outside his cell door, solid steel save for a peephole, began to wonder why the shower had been turned on for so long, finally rushed in to investigate, and discovered the most celebrated prisoner in federal custody crumpled, barely conscious, on the floor of the stall.

I had been denied further access to Valachi once he was in Milan, but an exception was now made in the hope that I might help dissuade him from trying to take his life again. When I saw him, he still had a deep, ugly red line running three-quarters of the way around his neck where the cord had left its mark. His left ankle was still quite swollen after catching the force of his fall. Both knees and his left shoulder were black and blue from slamming into the sides and floor of the stall.

Why would such a prize catch of the federal government, guarded night and day for his own protection, want to kill himself? The reason was as ugly as the red line on his neck. Valachi's entire adult life had been conditioned by the belief that the Cosa Nostra was invincible. Having openly defied it, however, the idea that his story would be published was final proof to him that he might have been wrong—that there were forces in this country immune to its power and influence. Now suddenly, from this point of view at least, none of this seemed to be so. On April 6, two weeks following his transfer from Washington to Milan, Valachi was told that the whole thing was off. For good measure he was also deprived of his precious hot plate, a small matter perhaps, but a perfect symbol for him of his abrupt fall from grace. Five days later he made his suicide try.

During my visit with him afterward—it was the last time I was allowed to see him—I found a bewildered man. "I don't understand," he said. "From doing right, all of a sudden I'm doing wrong."

The fact of Valachi's nearly total recall is undeniable. Except for minor memory lapses, his FBI interrogations checked out on every verifiable point. I can attest to this because I read the reports on these interrogations at the Westchester County Jail and Fort Monmouth exactly as they came into Washington before they were documented. And beyond the areas of interest to law enforcement officers, it is virtually impossible to fault Valachi on even the most obscure details in his story—the name of a policeman who arrested him in the 1920s, a tabloid headline in the early 1930s, the price on a race one of his horses won in the 1940s.

In this regard, I got my own comeuppance. During one of my interviews with Valachi, he launched into a brief account of a power struggle over the control of jukeboxes in Westchester County. As a result of this, a boyhood pal of Valachi's named Charles Lichtman, who was not in the Cosa Nostra, was forced out of the action. But for old time's sake, he decided to see what he could do for Lichtman—without success as it turned out. Since the episode did not seem terribly important, I was in the middle of changing the subject when Valachi tossed in a cryptic remark that his friend Lichtman had told Senator McClellan "all about me."

That evening I searched all through the McClellan subcommittee hearings on Valachi, but I could not find any mention of a "Lichtman." I was sure that at last I had caught Valachi in a flight of fancy, however inconsequential. Then, a few days later, I recalled that there had been another McClellan subcommittee investigation—into underworld influences in the labor-management field—which had taken place several years earlier. And there indeed, lost and forgotten in literally hundreds of thousands of words of testimony, was an appearance by one Charles Lichtman. He was questioned by none other than Robert Kennedy, at the time chief counsel for the subcommittee, on December 4, 1958— long before anyone ever heard of Valachi in his present role. It follows here in part:

KENNEDY: Now you still decided you wanted to get your jukebox union back, and did you go back up there?

LICHTMAN: I went back there a number of times, but I found out that Mr. Getlan had a pretty good hold on it because he had brought some mobsters into the picture.

KENNEDY: You talked to a man named Valachi?

LICHTMAN: Yes, sir.

KENNEDY: Who is Valachi?

LICHTMAN: I happened to know Valachi from around Harlem. He thought he could straighten it up for me.

KENNEDY: He is an associate of Anthony Strollo, alias "Tony Bender" and an associate of Vincent Mauro. He was convicted of violation of the Federal narcotics laws in

1956 and sentenced to five years. He has 17 arrests and 5 convictions. He told you he could straighten it out?

LICHTMAN: Yes, sir.

KENNEDY: What happened then?

LICHTMAN: So he had me go up to a bar on 180th Street and Southern Boulevard. I sat out front at the bar.

Kennedy: Who met at the bar?

LICHTMAN: Well, I met Getlan there and I saw this Blackie. Then Mr. Valachi came and they went into the backroom and they had a meeting.

KENNEDY: Who was in the backroom?

LICHTMAN: I don't know who else was there.

KENNEDY: Did you know Jimmy "Blue Eyes" Alo was in the backroom?

LICHTMAN: I didn't see him myself. I saw Tommy Milo.

KENNEDY: A notorious gangster in New York?

LICHTMAN: I imagine so.

KENNEDY: They had a meeting as to who was to control the jukebox union in Westchester?

LICHTMAN: Yes.

KENNEDY: What did they decide?

LICHTMAN: From what they told me, from what Valachi told me at the time, my partner Jimmy Caggiano took $500 and sold me out and for that reason I couldn't get anything back no more. You have no racket connections, so you are nobody. You are out.

KENNEDY: You had Mr. Valachi, who has a pretty good record. . . .

Five years later, as Attorney General, Kennedy would say of Valachi: "For the first time, an insider—a knowledgeable member of the racketeering hierarchy—has broken the underworld's code of silence. Valachi's disclosures are more important, however, for another reason. In working a jigsaw puzzle, each piece in place tells us something about the whole picture and enables us to see additional relationships. It is the same in the fight against organized crime. Valachi's information [adds] essential detail and brings the picture into sharper focus. It gives meaning to much that we already know. The picture is an ugly one. It shows what has been aptly described as a private government of organized crime, a government with an annual income of billions, resting on a base of human suffering and moral corrosion."

The information Valachi provided was primarily intelligence, although his testimony in April 1968 was crucial in the conviction of one of the fastest-rising young Cosa Nostra powers in Brooklyn, Carmine Persico, Jr.* By the time Valachi had been persuaded to talk, the statute of limitations had run out on practically all the specific crimes he discussed—except murder. And in New York State, where the murders which had involved him took place, a corroborating witness is required in the absence of physical evidence. Notably enough, in the two murder cases that offered the best prospects for a successful prosecution, an accomplice named

---

*In a very real sense Valachi was as much on trial as Persico. His testimony linking Persico to a $50,000 hijacking lasted about fifteen minutes. For days, however, a battery of defense lawyers questioned his competency as a witness. When this strategy completely backfired, they continued to hammer away at Valachi, endlessly trying to catch him in a contradiction about his own past in order to discredit his memory, then attempting to prove that he had been coached on what to say by the Justice Department, finally trying to portray him as an alcoholic or molester of little girls—all of which turned out to be equally futile. It only took the jury five hours to find Persico and his codefendants guilty. The case is being appealed.

by him mysteriously disappeared and is presumed dead after it became known that Valachi was talking.

How important was Valachi? According to William Hundley, operationally in charge of the Justice Department's drive against organized crime before, during, and after Valachi talked: "What he did is beyond measure. Before Valachi came along, we had no concrete evidence that anything like this actually existed. In the past we've heard that so-and-so was a syndicate man, and that was about all. Frankly, I always thought a lot of it was hogwash. But Valachi named names. He revealed what the structure was and how it operates. In a word, he showed us the face of the enemy."

Valachi was peculiarly equipped to do so. In the Cosa Nostra's paramilitary organization he was on the order of a master sergeant working out of headquarters, and there are few men still alive today who can match his thirty-three years of active duty. His service, moreover, exactly coincides with the birth and growth of the modern Cosa Nostra. He was, as they say, there when it happened; he knows where all the bodies are buried.

3

Italians did not invent organized crime, nor did they introduce it to this country. Indeed, when the first great wave of Italian immigrants arrived in the late nineteenth century, they found a flourishing underworld then primarily in the hands of the Irish and the Jews, although just about every ethnic group has had a crack at it at one time or another, including good old-fashioned American families like the James boys. But what a very small number of these Italian newcomers—mostly from Naples, Calabria, and Sicily—did bring was a traditional clannishness, contempt for lawful authority, and a talent for organization that would eventually enable them to dominate racketeering in the United States.

The first significant step in this direction occurred in the 1890s, when a gang of Sicilians gained control of the New Orleans waterfront. No cargo moved on or off the docks without their being paid tribute. Then the city's chief of police, probing too energetically into their activities, was murdered. Nineteen members of the gang were brought to trial, the case against them apparently airtight. But a dreary pattern, so familiar today, was already being set. The best criminal lawyers in the country were hired and, helped no end by some jury tampering, won acquittals for all but three of the defendants. The first time, however, the strategy backfired; after the verdict was in, an enraged crowd wound up lynching eleven of them and very nearly caused a diplomatic break between Washington and Rome.

But such a bold penetration of the established underworld was exceptional in those days. The earliest organized Italian criminals in America, the Black Hand extortion rings, preyed almost exclusively on the vast majority of their decent, hardworking countrymen who had settled here. The name came from a crudely drawn black hand on the bottom of a letter demanding money from a particular victim and usually threatening the death or mutilation of his children if it was not paid. With the memory of the dreaded Mafia in Sicily or the Neapolitan Camorra still fresh in their minds, distraught parents promptly forked over the cash. Black Hand extortionists became such a problem that in New York City a special police squad, led by Lieutenant Joseph Petrosino, was assigned to hunt them down. Furious that a handful of his own people was tarnishing the name of all Italians, he did his job too well; in 1909, while in Sicily to exchange criminal intelligence with local officials, he was shot in the back and killed.

Joseph Valachi was five years old when Petrosino died. He had been born on September 22, 1904, in Manhattan's East Harlem where remnants of a once-large Italian population still can be found. Both of Valachi's parents came from Naples. They had seventeen children, but only six survived; Valachi was the second oldest. His alcoholic father was initially a vegetable pushcart peddler and then a laborer on a garbage scow. Of his childhood Valachi recalls:

My mother's name was Marie Casale. She was about 5 feet 7 and heavyset until she got older, and then she dropped down to about 120 pounds. My father's name was Dominick. I'd say he was about the same height and weighed about 160 pounds, never dressed clean, and had a big mustache. My older brother, Anthony, the last I heard is still in the bughouse.* My kid brother, Johnny, was a drifter, and I couldn't do a thing with him. He was found dead in the street, and the cops claimed it was a hit-and-run accident. I heard that they pulled him in for questioning and worked him over too much. My three sisters all got married, so I won't talk about them.

My father was a hardworking man, but he drank too much and my mother always had a black eye. The neighborhood in East Harlem was pretty rough in those days, and you could hardly walk around without catching a bullet. I remember my father had to pay a dollar a week for "protection," or else his pushcart would be wrecked.

He would make pretty good money selling all kinds of vegetables, but like I said, he would drink it all up. When I was a little boy, I used to help him with the vegetables. One time I was pushing the cart, which only had two wheels in front, and I slipped. To make a long story short, I dumped the tomatoes all over the street, and my father beat the hell out of me. Later, when he went to work at the City of New York garbage dumps at 107th Street and the East River, I worked with him.

We were the poorest family on earth. Anyway that's how it seemed to me. When I was growing up, we lived in different places but always around East 108th Street. One apartment was at 312 East 108th

---

*State Hospital for the Criminally Insane, Dannemora, New York.

Street. I'll describe it for you. There were three rooms, no hot water, and no bath. The toilet was out in the hall. The only heat was from a stove in the kitchen. We would bring home wood and used coal that we got from the dumps; we stored it in the room me and my brothers slept in. It got so that the whole room would be stocked to the ceiling in the winter, and boy, was it dirty! For sheets my mother used old cement bags that she sewed together, so you can imagine how rough they were.

I dared not think of any girls. I feared that they would want to come into the house. If they did, I think I would have died—from shame, of course. There was one girl I liked who lived across the street. She lived on the top floor, and we were on the ground floor. She could look right into our house when the light was on, and when she told me that she saw me at night before going to bed, I used to get heart failure because I felt guilty about how filthy everything was. Lots of times I left my room in the middle of the night and sneaked into a stable down the block full of wagons. Why would I want to sleep in a wagon instead of my own bed? If you want the truth, it was to get away from all the bedbugs.

I did the best I could to keep myself clean, but with all the dust from the coal we had stored in the house, it wasn't easy. There was a public bath at 109th Street and Second Avenue. To get into the bath sometimes, the line would be a block and a half long. Then when you were in the baths, they would give you only so much time, and you had to get out of there. Believe me, it wasn't much time.

I was supposed to go to school, but to be honest about it, lots of times I didn't. I got picked up once by the truant officer; all I got was a warning. Then, when I was eleven, I hit a teacher in the eye with a rock. I didn't mean to do it; I was just trying to scare her. Anyway, I was sent to the New York Catholic Protectory, which is where the court sent you for being in trouble like I was, or if you were an orphan. It was in the Bronx, and it was pretty rough. As far as the brothers were concerned, some of them were okay and some were real bad. You wouldn't believe what some of them were like, fooling around with the young kids, but I don't want to go into that.

The roughest one was Brother Abel. He was in charge of the tailor-shop, and he would lay into us with his tape stick something awful. It didn't matter whether we did anything wrong or not. The best thing to do was stay out of his way unless you were looking for a beating. Then one day Brother Abel died. They put his body on display in the chapel. I'll never forget it. All the kids from the five yards of the protectory had to line up to view him and pay their last respects. All told, I'd say there were about 300 of us. I was near the end of the line, and when it was my turn to view the body, I almost fainted. Brother Abel's chest was all covered with spit, so what could I do? I spit on him, too.

I got out of the protectory when I was fourteen and went back to school for a little while. But when I was the age of fifteen, I got my working papers and went to work with my father at the garbage dump. At the end of the week my father would take my pay, too, and at night there would be war at home. This wasn't any good for me, so with one or two other fellows on the block, naturally I started to steal to have a little money of my own.

By the time he was eighteen, Valachi's petty thievery had led to full-fledged membership in a burglary gang, working out of East Harlem's 107th Street, called the Minute Men because of the speed of their operations. Valachi always had a great interest in cars, and his primary responsibility in the gang was to be the "wheelman," or driver. This gang pulled off literally hundreds of thefts between 1919 and 1923. Their methods were simple enough. A garbage can would be used to smash a store window, and whatever was inside, usually furs or jewelry, was taken out and sold to a fence. With police car radios not yet commonplace, the gang could count on several minutes before the law arrived from the nearest precinct house. Some merchants relied on a private alarm system provided by the Holmes Electric Protective Company to guard against thefts, but this did not appreciably close the time gap. Although Valachi's record shows that he was picked up five times on suspicion of burglary or larceny charges, he escaped a jail sentence until the spring of 1923:

The Minute Men were the talk of the underworld, even if it sounds like I'm exaggerating. We were real cowboys, meaning that we were riding high, and when we met other mobs in the cabarets around town, I must say everybody wanted to know who the wheelman was. The boys would point at me, and the other guys would always buy me a drink and say, "Well, good luck, kid." I must say it made me feel real good, as I was only nineteen at the time. It was the same when we went by Sadie Chink's place on Manhattan Avenue. She always had five or six girls in the house, and it was safe there because Sadie was paying off the cops. Boy, when one of the girls said, "Gee, I've heard about you," it made you feel like a real knock-around guy. They used to charge $5, but we always gave them $10. They were the only girls I saw because I didn't care to settle down. It was no time to start falling in love and worrying about a family.

I must explain that we were called Minute Men because we always got away in a minute's time or less. This gave us all the time we needed. Even the store that had what we called Holmes protection took from five to seven minutes to bring anybody. I know because I threw a brick through a store window on 125th Street to see how long it would take. The only thing we had to worry about was if the bulls were already in the neighborhood.

That's what finally happened on this job on Tremont Avenue and

177th Street in the Bronx. It was a store full of silk. We had been look-ing it over for a couple of weeks, and it looked like a cinch. It didn't even have a Holmes alarm system. We were in a Packard touring car when I pulled up in front of the store on the Tremont Avenue side. I for-get what we used to break the window—I think it was a milk can. Any-way, one of the fellows went into the store to start handing out the bolts of silk. Two others were carrying them out to the car. The fourth one was standing on the corner so he could see down 177th Street. All of a sudden the guy on the corner comes running over to me, saying that there was a car coming down the side street. As he is telling me this, I see another car coming slowly down the avenue towards us, and I tapped the horn to warn the fellows in the store to get out. Then, just as they were all back in the Packard, I see another car creeping up right behind us. Gee, I said to myself, somebody must have tipped the bulls. All of a sudden I felt a gun against my head. It was a cop. I found out later it was this Captain Stetter who had been after us. He said, "I finally got you after three years."

Of course, I kept the motor in the Packard running, and this was where all my practice learning how to drive came in handy. When Captain Stet-ter told me to take the car out of gear, I said, "Okay," and pretended to start getting out of the car. But I was doing this to turn the steering wheel so I could get the car away from the curb and out onto the avenue. All at once I dove down on the floor under the dashboard, got my right hand on the gas pedal, and held the wheel with my left hand. You must understand that all this was done in a couple of seconds. I pushed down on the pedal as hard as I could with my hand and took off. Naturally the cops around us started shooting, and all the fellows ducked down. They shot the whole windshield off. Believe me, there wasn't any glass left in it.

When I looked up, I saw I was in the middle of Tremont Avenue. All

I could think of was to get the hell out of there. The cops had been getting some new Cadillacs, and one of them was about two blocks behind
us. Then what do you think happened? Just as I was coming to a corner, a trolley car pulled out and stopped right in the middle of the
avenue. I don't know what to do. If it starts to go forward again, we are
dead, as I was going too fast to make a right turn. If it stays where it
is, I could make it to the left. But to do this, I have to go across the
sidewalk, jumping the curb. Well, the trolley stayed still, and that was
all I needed, so I did it.

By now I'm hitting eighty miles an hour, which was a lot of speed in
those days, and the cops are way behind me. They had to slow up when
they saw the trolley car. For the first time I see some blood running
down my wrist and I realize that I am hit in the arm, but I figured this
isn't the place to stop and look at it. I turned down the Grand Concourse
and kept heading for Harlem. At that time they had police booths along
the Concourse, and as we came tearing down it, they blazed away at us.
Don't forget we were in a touring car, and as we had been speeding all
this time, the top finally flew up and was hanging over the back end.
When we got to the bridge into Manhattan, I kept going like a wild man,
and we lost them. I was only laid up a few days with my arm since the
bullet missed the bone and came out the other side. But the bulls
traced the car even though we had bent the license plate to hide the
number—I still remember it, 719864—and pulled me in.*

They kept me in the Bronx County Jail for about six weeks. Then I
went to court and pleaded guilty to attempted burglary. When I went to
get sentenced, the owner of the store was there. After my name and

---

*In the early 1920s, with few automobiles on the street at night, Valachi rarely
used one that was stolen. Somebody his age driving around in a car often
would be stopped by the police simply to check the registration papers.

the charge was called out, he jumped up and started yelling at the judge that he wanted to know where the "attempt" was when he was out $10,000 worth of silk. I must admit that he had something. But you see I was learning about the law, and believe me there was a lot to learn. When you plead guilty, you can plead to a lesser charge. But the owner of the store wouldn't buy this. He said it wasn't fair, and where was his silk? My lawyer, who was Dave Goldstein, told me to answer.

So I said, "I threw it in an empty lot."

With that I was remanded for a couple of weeks while they worked the whole thing out about whether I had to withdraw my plea and stand trial. The decision was that the court could not force a defendant to do this, so now I was up for sentencing. Because I was under twenty-one, the judge could have sentenced me to the Elmira Reformatory for eighteen months. But my lawyer told me to go in front of him with a strut. That way he would figure me for a tough guy and send me to Sing Sing Prison, where I would be out in only nine months.

This is how it worked. At the reformatory I wouldn't get any time off for good behavior. But my sentence at Sing Sing would be one year and three months to two years and six months. With time off I will only do eleven months and twenty days before I'm up for parole. I will get credit for the time I was in the Bronx County Jail already; if I watch my step, I will be out of Sing Sing in less than nine months.

When I heard this, I naturally went in front of the judge with a strut. I could see right away that he didn't like the way I was acting. I remember just what he said, "Do you think you're fooling me acting like a tough guy? Well, I'll tell you something. I'm going to send you where you want to go. You know why? Because the sooner you're out, the sooner you'll be in front of me again."

Well, who cared what he said?

A few days later Valachi was taken to Sing Sing. At first his strategy to escape a longer sentence in a reformatory seemed a dreadful mistake. He was placed in a tiny, clammy cell with no sanitary facilities other than a bucket. To his relief, however, he learned that he would stay there for only ten days while he was being processed. And even before the ten days were up, he was rather enjoying himself. Taken to a special prison showing of a Broadway musical—*The Plantation Review*—he recalls, "I had such a great time I couldn't believe that I was at Sing Sing."

He was then assigned to a construction gang building a new cell-block, delighted to be out of his dungeonlike quarters, and life at Sing Sing passed quickly and uneventfully for him. Better yet, he was released on parole after serving his nine months and returned to New York, where, within days, he was back "crashing" store windows.

There had been some changes during his absence. His old 107th Street burglary gang had now taken to hanging out on 116th Street "where all the action was." Valachi found the new life there exhilarating, especially at night at the Venezia Restaurant:

Guys were coming there from all over the city. Besides us Italians, there were the Diamond brothers, Legs and his brother Eddie, there were other Jew boys, and Irish guys from down around Yorkville. Sometimes you saw Lepke and Gurrah and also Little Augie from the East Side downtown.* But the big man on 116th Street was Ciro Terranova,

---

*Louis (Lepke) Buchalter, Jacob (Gurrah) Shapiro, and Jacob (Little Augie) Orgen were among the most feared Jewish mobsters of the day. Shapiro got his odd alias as a young hoodlum, when pushcart peddlers would cry, "Get out of here, Jake's coming!" The first part of the frenzied warning sounded more like "Gurrah," and the name stuck.

the Artichoke King. He got the name because he tied up all the artichokes in the city. The way I understand it he would buy all the artichokes that came into New York. I didn't know where they all came
from, but I know he was buying them all out. Being artichokes, they
hold; they can keep. Then Ciro would make his own price, and as you
know, Italians got to have artichokes to eat.

To his dismay, however, Valachi learned that the new 116th
Street gang had acquired a new wheelman during his absence, and
now they had no place for him. He put together a little group of
his own and went on a series of petty thefts until he had enough
money to buy a 1921 Packard under an assumed name. The car
enabled Valachi to embark on more lucrative and sophisticated
operations, using homemade tools to jimmy open shop and loft
doors instead of risking the noise of breaking windows.

His new confidence nearly did him in. While he was trying to
force his way into a Bronx warehouse full of furs, one of the tools
snapped. Another member of the gang suggested returning to East
Harlem for a replacement. "Sure," Valachi said, "why not? Get
into the car." Just as they were ready to depart, a patrolman walking his beat fired at them. The slug lodged in the base of Valachi's
skull. "All I heard was that one shot," he says. "Next I heard
someone yell. 'He's dead! What'll we do with him?'" Then Valachi
passed out. His companions dropped his body off at 114th Street
near the East River, fired six shots in the air, and left in the hope
that the police would consider it a gangland murder that had no
connection with the attempted robbery. But when they passed by
an hour later, Valachi was still lying in the street. Finally they
noticed he was still breathing and brought him to a neighborhood
doctor noted for his cooperation in such matters. A bottle of

Scotch was used as an anesthetic to remove the bullet. Valachi remembers coming to in the doctor's office and being told, "Drink this." But his proudest moment was when he heard the doctor say through his haze of pain, "This kid won't die. He's built like a bull." It took him two months to recover. "The thing that saved me," he says, "was all that work I did with the sledgehammer at Sing Sing. I was in real great shape."

But his near-fatal wound was just the first episode in a star-crossed period for Valachi. Soon after he had returned to robbery as a regular way of life, he fell in with a man who would have a profound influence on his future, Dominick (The Gap) Petrilli. The Gap suggested a joint venture that seemed ideal—a loft in upper Manhattan crammed with silk, without an alarm system. Valachi agreed at once, but his Packard wasn't big enough to carry off the booty. Thus another friend, Joseph (Pip the Blind) Gagliano, who had his own car, was recruited. Two more hoodlums rounded out the group. Everything was proceeding smoothly, most of the silk successfully placed in the two cars, when they spotted a figure crouched in a corner of the loft next to a telephone. He turned out to be the night watchman. Outraged at this unexpected discovery, two of the gang attacked him with iron pipes. As the watchman tried to ward off the blows, he screamed that he had already called the police. But when everyone ran outside, Valachi's Packard would not start. "Well, there goes another car," he said as he jumped into Pip the Blind's Lincoln, barely escaping the police. The next day a member of the gang came to Valachi's house and asked him for the keys to retrieve the Packard, arguing that it was ridiculous to lose it. Valachi stupidly agreed to the request. "You got to remember I was just getting over that bullet in the head," he says, "and I wasn't thinking too good." Worse yet, the friend took Valachi's youngest sister along for the ride. The police, of course, had the car staked out, followed the couple home, and the next thing Valachi knew he was under arrest again.

Released on bail, he was headed for still more trouble. He had become increasingly friendly with some Irish toughs during his free nights at the Venezia Restaurant, and to his amazement one evening, when he returned home from visiting a friend in another part of the city, he was suddenly accosted by an angry group of Italian youths whom he had grown up with. "What the hell's going on?" he demanded.

"You know what happened on 116th Street," one of them, Vincent Rao, snarled. "Those Irish guys shot up the block this morning, and you were driving the car."

Valachi finally convinced Rao that he had been nowhere in the area. Then Rao explained that an automobile had suddenly roared down the street spewing bullets and that Valachi's Irish friends had been recognized. Someone also thought that he had seen Valachi behind the wheel. Apologizing for the mistake, Rao took Valachi aside and said, "Anyway, these Irish like you. I want you to set them up for us. Will you do it?"

"I'll think about it," Valachi replied.

The next day, typically suspicious and wary, he decided to approach one of his Irish friends, known to him only as Mike, to see if there was another side to the story. No sooner did he confront Mike than he found himself covered by a gun and was told, "Let's go see the others."

As suspect with them as he had been with Rao the day before, Valachi was told by the Irish that they had not initiated the shooting incident. "Okay," he said, "that's good enough for me. There's no reason for you to lie. I want to join you guys. I like your style. You got a lot of guts."

"I don't know," one of them said. "After all, you were brought up with them other guys."

"You're right," Valachi replied, "but I ain't sticking up for you. I'm sticking up for my own principles. I'm supposed to double-cross you guys, and you ain't done nothing to me. So why should I set you up for them?"

That night, in a temper as characteristic as his caution, Valachi telephoned Rao. "After this," he said, "when you people meet me, shoot me because I am out to shoot you guys."

"What's the matter with you?" Rao asked.

"What do you mean, what's the matter? First of all, don't I have enough troubles of my own? I got enough things on my mind being out on bail. So all of a sudden everybody is looking to shoot me without any reason. And you started it, asking me to do something that only a dog would do. Why are you picking on me to pick on those guys? I don't like it. So watch out!"

"Jesus," Rao said. "I want to talk to you."

"Fuck you!" Valachi yelled and hung up.

The Irish gang, as Valachi called it, was in fact a polyglot out-fit that also included two Jews and, counting Valachi, even three Italians. For about three months intermittent gunplay with the 116th Street gang took place. The only casualties were two inno-cent bystanders—"Poor souls," as Valachi puts it—who happened to get in the way of stray bullets. Meanwhile, Valachi had begun to regret the "guts" he had originally admired in his new compan-ions. Instead of the burglaries he was used to, they favored armed robberies embracing everything from subway stations to bank shipments of cash. "They had a lot of nerve," Valachi recalls, "but no business sense. I figured it would be a short life, staying with them." He tried to persuade them to adopt his operational methods, breaking into warehouses and lofts at night. Finally they agreed to let him lead them in the looting of a clothing store. After picking the lock, Valachi took most of the gang inside, while stationing two others on the sidewalk as lookouts. But when he returned with an armful of suits, he discovered that his two lookouts, doubtless bored with guard duty, had lined up half a dozen passersby against the wall at gunpoint and were removing their wallets. At once Valachi called a halt to the operation and drove off. "Jesus," he said to the two lookouts. "What are you doing with those people? This ain't no game. You do that and it ain't a burglary anymore; it's a stickup. Them people can identify us."

"Well," he was told, "we don't like this kind of work."

Aware that he could never change their ways, Valachi began to think about leaving them. Any doubts he might have entertained vanished forever when one member of the gang was wounded during a holdup. "By the time we got him to a doctor," he recalls, "this guy was in bad shape. As I understood it, he was hit in the lung, and it was a good thing it was winter, as it was so cold the blood kind of froze in him. If it didn't freeze, he would have bled to death."

Valachi's predicament was resolved when a truce between the Irish gang and his former cohorts on 116th Street was arranged by the Artichoke King, Ciro Terranova. Ostensibly, this got Valachi off the hook, but within days he was brought to trial on the loft theft in which his car had been traced by the police. He was convicted and was returned to Sing Sing to serve not only his new sentence, but the remaining time on his first offense as well—a total of three years and eight months.

Valachi's arrival there was something of a reunion. "It was my second time," he notes. "I knew most of the inmates and they wanted to know, first of all, how much I was in for. You got to understand that the guys that I was close with in Sing Sing were all Italian, and I was surprised they welcomed me and didn't have no hard feelings because I went with the Irish." His relief, however, was short-lived:

I'm at Sing Sing, all settled in, when I read in the papers that one of the Italian members of our gang, a kid named Frank LaPluma, got killed. They shot him sitting on a stoop one morning. The way I made it out, it didn't make no sense. Well, all I could do was wonder what was going on. Then this other guy who was in with us, Dutch Hogey, comes up on a twenty-year rap. See what happens when you use guns all the time. He had hit a cop and got wounded himself. I went up to the prison hospital to see him and we talked old times. I told him, "You should've listened to me. Pete Hessler is already in the death house."

"I know," he said, "you're right, Joe." Then the Dutchman tells me what I can't believe. "They sold you out," he said. I said I didn't know what he was talking about. "I mean they made peace," he said, "on the condition that you and Frank must die. Ciro Terranova fixed the whole thing." Then the Dutchman told me to watch myself. He said that if they got one, they'll get the other—meaning me. But I figured I was safe enough where I was.

Right after this I was mopping up one day in the dormitory with Dolly Dimples. His real name was Carmine Clementi, but he was known as Dolly Dimples because he was a handsome kid with blue eyes and light brown hair. There were two singers in those days called the Dolly Sisters who were very big, and everyone at Sing Sing used to call us the Dolly Sisters because we went around together all the time. Anyway, to get back to the dormitory, Dolly had gone off somewhere, and another guy who was helping to clean up, his name was Angelo, was in the toilet.

Just then there was a knock on the dormitory door, and a kid by the name of Pete LaTempa said he wanted to get something from under his bed. I didn't think anything about it. I knew this LaTempa. He had come up the river after I did, but I never had much to do with him, so I let him in and went about my business with the mop.

All of a sudden I felt sort of a sting—that's the best I can describe it—under my left arm. I looked behind me, and I saw this LaTempa with a knife in his hand. By now Angelo had come out of the toilet and was standing there, looking at me with his eyes bugging out. He was trying to tell me that I was cut, but he was stuttering so much that I couldn't make out what he was saying. I put my hand down under my arm where he was pointing, and I kind of felt it going right inside me. Then I saw all the blood. Believe me, it was all over the place. So naturally I went after LaTempa, and he started to yell how bad I was cut, hoping I would forget him and worry about myself. But I just kept going, and when I caught him, I let him have a couple of good raps on the mouth. He was smaller than me, and I would have killed him with my bare hands, but by this time my knees were getting weak, and he ran out.

Maybe after a minute Dolly came back. If he had made it sooner, LaTempa would never have gotten away. I was down on the floor bleeding worse than ever. What saved me was that the hospital was only one flight above the dormitory. Dolly carried me up there. He said, "Don't worry, I'll get the son of a bitch." In the hospital they kept asking me, "Who cut you, Joe? Who cut you?"

I said I didn't know.

When they finished sewing me up, I had thirty-eight stitches running from right under my heart and around to my back. I still got the scar.

A few days later LaTempa, apparently fearful of retribution from Valachi, turned himself in to Sing Sing officials and was transferred to another penitentiary. No more attempts were made on Valachi's life during the rest of his term. In the prison school he finally completed the seventh grade, learned how to read and, however crudely, to write.

But nothing left a greater impact on him than a series of talks he had with an old-timer named Alessandro Vollero, one of the most prominent of the early Italian gangsters in Brooklyn, who was serving a life sentence for murder. What especially attracted Valachi to him was that his victim was a brother of Ciro Terranova. From Vollero he got some idea of the deep-seated hatreds that raged then in the Italian underworld between its two chief elements, the Sicilians and the Neapolitans.* Vollero told Valachi, "If there is one thing that we who are from Naples must always remember, it is that if you hang out with a Sicilian for twenty years and you have trouble with another Sicilian, the Sicilian that you hung out with all that time will turn on you. In other words, you can never trust them. Talk to me just before you get out of here, and I will send you to a Neapolitan. His name is Capone. He's from Brooklyn, but he's in Chicago now."

From Vollero, Valachi also received his first inkling of a secret criminal society he would one day know as the Cosa Nostra. But when Valachi tried to delve further into it, Vollero simply said, "Take it easy, kid. You'll learn all there is to know in good time. It's not for me to say it."

---

*Vollero was a member of the Camorra, the Neapolitan version of the Sicilian Mafia. Vollero's trial in 1918 was a sensation of the day. In court it was revealed that his gang had a standard toast: "Health to all Neapolitans and death and destruction to all Sicilians!"

Valachi was released from his second go-around at Sing Sing on June 15, 1928. He looked on the experience as not without advantages. "I came home," he says, "with an education. I didn't learn much in that school, but at least I could read something and know what I was reading. Before I went back to Sing Sing, I could hardly make out the street signs. But the real education I got was being worldly-wise. I could sit here all day and how am I going to explain what I mean? It's just all the things you learn about human nature in another world, and believe me, Sing Sing was another world."

Valachi faced some immediate problems when he was released. Even if he had wanted to, he could not return to the Irish gang. His old worry that their ways would do little for his life expectancy had clearly come to pass; of its seven members, one hanged himself in prison, a second had simply disappeared, a third had been electrocuted for killing a police officer, the fourth, Frank LaPluma, had been murdered on Terranova's orders, and two others had ended up killing each other in a drunken brawl. Only Dutch Hogey, in Sing Sing, and Valachi still survived.

He could have opted for Chicago, as Vollero had suggested, and joined Capone, but the prospect of leaving the familiar surroundings in East Harlem held little appeal for him. Having decided to stay, however, it was vital for Valachi to resolve his situation with Ciro Terranova, Vincent Rao, and the rest of the 116th Street group. He had some reason for hope. Shortly after the knife attack in Sing Sing, his friend The Gap was sent up on a robbery conviction, and Valachi sought his advice. "I don't think you have to worry," The Gap said. "You know, time heals everything. Just play dumb and keep to yourself for a while. A lot of people think you got a bum rap from the boys."

Remembering this, he now approached Frank Livorsi, Terranova's chauffeur and bodyguard. Valachi told him what The Gap had said and added, "I just got out. See if you can find out what's what for me."

A few days later Livorsi reported, "You mind your business, and everybody else will mind theirs. There are no hard feelings. What's over is over."

This stopped far short, however, of an invitation to rejoin his boyhood pals in the 116th Street gang, a number of whom had moved up into racketeering.* "In other words," as Valachi notes, "they were no longer stealing. They were mobbed up. But as I am not with them anymore, I don't know exactly what they are doing. I have to hustle for myself. Well, I figure if it has to be, it has to be."

Left to his own devices, Valachi returned to the one occupation he knew—burglary—and put together a gang of six young Italians. They operated two or three nights a week, and their total "swag," as he calls it, averaged about $1,500 a week. "It didn't leave so much when we split it up," he says, "but it kept us going, existing, and that was all I could ask."

At this point, in early 1929, The Gap returned from Sing Sing. "The Gap," Valachi says, "was like my schoolteacher, only in crime." Valachi found everything about him admirable—his expensive clothes, his insistence on always picking up a check, his toughness, his "personality"—so he went to him for help in

---

*The word "racketeering" has come to be applied to a number of criminal operations whose common denominator is extortion of one kind or another. Its origin is somewhat obscure. According to the most popular theory, it stems from late-nineteenth-century neighborhood get-togethers given by New York City political clubs—affairs that were called rackets because of their boisterous nature. Local gangs then began sponsoring their own rackets, for which merchants in the area had to buy tickets or suffer the consequences.

improving his own status. "I knew The Gap was mobbed up," he says. "When I asked him how I could get in, he told me I would have to wait. He said to sit tight and keep in touch, and in time everything would be all right."

Buoyed by The Gap's promise to move him eventually into the rackets, Valachi continued to run his burglary ring. Money pressures, meanwhile, continued to build up. He had taken a separate apartment next to that of his parents. But he was their main source of support, since his older brother, an inept loner, had been sentenced to life imprisonment as a fourth offender for armed robbery.

Then his father died. Valachi had found him moping in his apartment one afternoon and grandly gave him $10 to get himself some wine. By the time The Gap stopped by to pick him up for a night on the town, the old man had already passed out on the floor. The two friends, laughing at the sight, simply put him to bed and went out. When Valachi returned later, he found his father in exactly the same position, except for a curious foaming around his mouth. Scared now, he did the unthinkable. He ran to the neighborhood cop and asked him to call an ambulance. But it was too late. His father never recovered from his alcoholic coma and was buried at the age of fifty-two.

Valachi had also taken his first mistress, a hostess he will only identify as May—"What the hell, she's married now"—who worked at a dance hall he frequented on 125th Street called the Rainbow Gardens. "She was my first steady girl," he says. "She was Jewish, but who cared about that? She was pretty, and she had a million-dollar personality. I remember she wore a size twelve, so when we burglarized a dress factory, I always held out a few instead of giving everything to Fats West, the fence."

Setting up housekeeping with a woman came as a surprise to young Valachi. He had, he recalls, been going out with May for about six weeks when she suddenly informed him that she had located an apartment in the Bronx for them and had already ordered the furniture. He made one feeble attempt to back out, claiming that he did not have a steady income, but May replied in classic fashion that she would continue working. "So I said okay," Valachi recalls. "If something went wrong, I figured she could always go back to her mother."

Life with May and the death of his father combined to have a salutary effect on the uneasy relationship Valachi had maintained with the Italian underworld ever since the days of the Irish gang. May's closest girlfriend was the mistress of Frank Livorsi, who had smoothed things over for Valachi after his release from Sing Sing. And as the two couples began to party regularly in the Bronx apartment, Livorsi even began to talk about completely patching things up with Ciro Terranova. Valachi was dubious about the prospects as much because of his own hatred for Terranova as anything else.

His father's death was equally effective in putting Valachi in contact with his old friends. Several of them came by to pay their respects at the wake, and after the funeral still more of them began to greet him cordially. Then he had a fateful encounter:

Through one guy, you naturally meet another. And now The Gap introduces me to Bobby Doyle. His real name was Girolamo Santucci, but he was a fighter once, and his ring name was Bobby Doyle. In those days a lot of Italian fighters used Irish names. I guess it was because they were easier to say. You could tell Bobby Doyle was an ex-fighter by the way his nose was flattened. He was very polite, a slow speaker, always beating around the bush. One day he asks me, "How are those guys you steal with?" and all that kind of stuff. Then he wanted to know if I could get six or seven good boys together in a hurry. I said, "I can get a dozen of them." Then he asks me how Frank, meaning Frank Livorsi, stands with me. I say, "He is the best. Didn't he straighten out everything for me when I got out of Sing Sing and no one would touch me? Why?"

"Nothing," he says. "I just want to know." But I can tell from the way he talks that he don't think too much of Frank.

So naturally I go to The Gap and tell him I don't like the setup. I'm thinking of everything Alessandro Vollero was telling me in Sing Sing about the Sicilians and the Neapolitans, and Bobby Doyle is Sicilian. But The Gap tells me, "You are crazy. Things aren't like they used to be. The feelings between Sicilians and Neapolitans is all past!" In other words, he is saying that everyone is mixed up now.

Well, Alessandro was right in his way, but I didn't know it at the time. Actually, in their hearts it was still the same, but it didn't show right out. They favored one another, but it wasn't broad, out in the open so you could see it, like in the days of Alessandro. Anyway, I said to The Gap, "I don't know what I'm getting into. It seems like every time I pick a side I end up the fall guy. Remember the Irish? Look at all the trouble I got into with them." The Gap says, "If I was around at that time, you wouldn't have been in no trouble. So take it easy. Bobby Doyle and me are vouching for you. That's all you have to know."

Well, who knows what the hell's going on? Some more time passes, and all of a sudden Bobby tells me to go see Tom Gagliano. I know he is a big shot, but just what or how I don't know. He is a big tall guy, a little bald. He looked like a businessman, and he was in construction work, for one thing. I first met him through the bouncer at this place on 116th Street. Me and Nicky Padovano, who was burglarizing with me, had worked over a couple of guys in one of the building unions for giving him some trouble. He liked me because I wouldn't take any pay for it. The reason I wouldn't take money is because I wanted to be recognized as a friend. I'm looking to meet people to get out of stealing, and you're not going to meet people by getting paid off. The thing is to be known as a man.

This day I see Tom Gagliano he says, "There's some trouble in the air, and I'm sure it's with people you don't like." I say, "Who's the trouble with?" And he mentions guys like Ciro Terranova. That's all I have to know, so I say, "Count me in." He said it was a million to one shot we don't make it, but if we make it, we're all right. He asked me point-blank if I would shoot somebody if they asked me to. I say, "Would your guys do the same for me?" He says, "Yes," and I say, "Yes," and that was all that was said.

Well, I'm in something, I still don't know what. But I got to get out of stealing. It's getting tougher all the time. They were talking about bringing radios in the police cars, and the traffic lights, which used to go out at three o'clock in the morning, are on all night, so that's against me, too. After all, if I'm being chased by one cop's car and I go through these lights, I will draw attention and have a hundred cars chasing me. Jesus!

Naturally, if I'm against Ciro Terranova, I got to be against my friend Frank Livorsi, too. So sure enough, a couple of days later the first contract I get from Bobby Doyle is to kill Frank. I said, "No, I won't do it." I run over to The Gap's house. I told The Gap how I pulled with Frank and Frank pulled with me. The Gap said he would straighten out everything with Bobby. The three of us meet at the Rainbow Gardens, and The Gap explains my feelings about Frank Livorsi and that I'm not going to take the contract. After a while Bobby Doyle says he understands, but then he takes The Gap aside and talks to him so I can't hear. I don't like the looks of this, and when The Gap comes back, I tell him so. The Gap says, "Don't worry. Bobby is just afraid you're going to run and tell Frank. I told him I would be responsible for you."

But I got to let Frank know something. So I told him if I ever move from the Bronx apartment, it will mean I've been approached. He knew what this meant, and after The Gap fixed everything about Frank, I did move out. If we saw each other on the street during all the trouble that followed, we would just wave and go the other way.

I wasn't going to dump no friend.

4

All the "trouble in the air," as Valachi would subsequently discover, was an enormous convulsion in the Italian underworld in 1930 called the Castellammarese War. While law enforcement officials were aware of its historical importance, they by no means had a complete picture of what actually was at stake. Valachi furnished the details at long last. It is a fascinating glimpse into all the savagery, avarice, and torturous double-dealing that still pervades the Cosa Nostra today despite its increasing sophistication and vaunted togetherness.

Organized Italian racketeering really did not begin to be a national force until the 1920s. Prohibition, of course, was the catalyst. In addition to those old standbys—prostitution and gambling—there was now a new illicit commodity that millions of Americans craved: alcohol. And it brought the racketeer riches and respectability beyond his wildest dreams; in effect most of the nation became his accomplice. The entire underworld, then monopolized by the Irish, Jews, and, to a lesser extent, Poles, cashed in on the Prohibition bonanza. But for Italian racketeers, especially, it was a chance at last to move into the big time. Bootlegging was something they knew about. For years, Prohibition or not, thousands of home distilleries had been operating in the ghettolike neighborhoods that Italian immigrants, like other ethnic groups before them, tended to crowd into after landing in this country. Thus they had a running start in the huge—and thirsty—market that had opened up, and from then on they bowed to no one.

By the end of the decade, despite the latter-day publicity given to Alphonse (Scarface Al) Capone, a vain, chunky little man named Giuseppe (Joe the Boss) Masseria had emerged as the most powerful single figure in Italian crime. Allied with him, besides Capone, was an awesome collection of mobsters of future note, including Charley (Lucky) Luciano, Vito (Don Vito) Genovese, William (Willie Moore) Moretti, Joseph (Joe Adonis) Doto, and Francesco (Frank Costello) Castiglia.

As strong as Masseria was, however, there were still a number of feudal aspects to what would evolve into the Cosa Nostra, and on top of the traditional tensions that existed between Sicilians and Neapolitans, the organization was fragmented by those who grouped themselves according to the particular village or region in Sicily or southern Italy from which they had emigrated. None was larger or more clannish than the Castellammarese, a contingent of men from in and around the Sicilian town of Castellammare del Golfo, who had settled in America as far west as Chicago. Even though dispersed, they remained closely knit under their chief in New York City, Salvatore Maranzano.

In 1930, bidding for absolute supremacy in the Italian under-world, Joe the Boss set out to eliminate Maranzano, as well as such other Castellammarese powers as Joseph (Joe Bananas) Bonanno and Joseph Profaci in Brooklyn, Buffalo's Stefano Magaddino, and Joseph Aiello in Chicago. And in all likelihood he would have succeeded had he not committed a fatal error in tactics.

But at the same time he was moving against the Castellammarese, Masseria tried to muscle in on one of his own gang leaders, Gaetano Reina, whose daughter Valachi would eventually marry and who then controlled most of the ice distribution in New York City—an enormously lucrative racket in the days before electric refrigeration. When Reina resisted, Masseria had him killed.

(New York City police records show that on February 26, 1930, at 8:10 P.M. one Gaetano [Tom] Reina, male, white, forty, of 3183 Rochambeau Avenue, Bronx, while leaving the premises of 1522 Sheridan Avenue, was shot and killed by an unknown male who fired both barrels of a sawed-off shotgun into Reina's body, causing death.)

Reina's murder would eventually unite his gang with the Maranzano forces against Joe the Boss Masseria. It also led directly to Valachi's recruitment into the Cosa Nostra. Masseria had replaced Reina with a man subservient to him named Joseph Pinzolo. While Reina's old lieutenants—among them Tom Gagliano, Thomas Lucchese, and Dominick (The Gap) Petrilli—outwardly accepted the appointment of Pinzolo, they were secretly plotting his overthrow.

"At the time," Valachi says, "I don't know nothing about all this, but the idea was to bring in new members like me that they could count on to help them. All I know is one day The Gap comes to me and says I got to meet this Joe Pinzolo, who is the new guy in charge of the mob. The Gap says he wants to look me over."

Valachi took an instant dislike to Pinzolo. He describes him as an ugly "greaseball"—the common Cosa Nostra term for older members not born in the United States—with a fat belly, a flowing handlebar mustache, and a distinct aroma of garlic about him. Matters did not improve much when Pinzolo said to him, "Hey, I hear you know all the girls up at the Rainbow Gardens. Call a couple of them."

Valachi, furious at being given what he considered was the role of a pimp, went to a telephone booth and pretended to make the call. After a minute or so The Gap came over and asked, "Are you getting the girls?"

"Are you out of your mind?" Valachi shot back. "Are you crazy? I have respect for those girls. If they come down here and see a guy like that, they'll faint. If these are the kind of people we are dealing with, I'm stopping right now."

"Shut up," The Gap whispered. "This guy is going to die. He is like a lot of other guys we are going to get rid of." Valachi returned to Pinzolo and explained that things were so busy at the Rainbow Gardens that none of the hostesses could leave. Then Pinzolo told Valachi to drive him to a midtown hotel. It was the last time Valachi ever saw him.

(According to New York police records, at about 9 P.M. on September 9, 1930, the body of Joseph Pinzolo was found lying on the floor of the interoffice of Suite 1007 in the Brokaw Building, 1487 Broadway, New York City. Suite 1007 was occupied by the California Dry Fruit Importers, leased to one Thomas Lucchese. Cause of death: gunshot wounds of left chest and neck.)

Although Lucchese was indicted for the murder, the charge was eventually dropped. The actual killer, according to Valachi, was Bobby Doyle. "I got the break of my life," he quotes Doyle as saying. "I caught him alone in the office."

Salvatore Maranzano and his Castellammarese were also striking back at Masseria. Their target was a vicious Masseria enforcer, Peter Morello, alias The Clutching Hand. His killer, Valachi learned later, was a Chicago-based gunman, whom Valachi knew only as "Buster from Chicago"; he looked like "a college boy" and carried, in the grand tradition, a machine gun in a violin case. "Buster told me," Valachi says, "that this Morello was tough. He said he kept running around the office, and Buster had to give him a couple of more shots before he went down. He said there was some other guy in the office, so he took him, too."

(Case No. 1226 of the 23d Precinct of New York City states that about 3:30 P.M. on August 15, 1930, Pietro Morello, 1115 Arcadian Way, Palisades, New Jersey, age fifty, was shot and killed in his office at 362 East 116th Street, by persons unknown. Cause of death: multiple gunshot wounds. Also killed was a visitor to the office, one Giuseppe Pariano.)

The two assassinations—by Maranzano's men and by the Reina loyalists now led by Tom Gagliano—were carried out independently of each other. "The way they explained it to me," says Valachi, "is that when Morello got his, the Gagliano people knew they didn't do it. So naturally they figure somebody else is in trouble with Joe the Boss. Then they found out that it was Salvatore Maranzano's doing." This brought the two dissident groups together under Maranzano's overall command. To seal their pact against Masseria, still unaware of the alliance, they agreed to kill a top Masseria henchman together. It was the first time Valachi was involved in a contract, and he vividly recalls the details:

At the time I'm just "proposed"—meaning I'm in line to be a member, but I ain't one yet. In other words, they are looking me over to see how I do. They tell me to rent this apartment up on Pelham Parkway because they found out this was the address of one of the guys under Masseria. His name was a hard one to say—Ferrigno. They wanted me to get the apartment because these other guys don't know who I was. Now the first thing I want to know is are they going to shoot this Ferrigno from the apartment. They tell me no, it's just to spot him, that they'll get him outside from a car, and I'll be driving the car.

Well, I move out of the place May and me have. I take the furniture except for one of the bedroom sets—we have two bedrooms there—which I give to May on account of I know she don't like the stuff she has in the bedroom at her mother's.

Now I'm in this apartment on Pelham Parkway. It's on the second floor. It's over a court, and on the other side of the court is the entrance to the apartment of Ferrigno. This way we can see him coming and going. One of the guys who stays on and off in my apartment is Joe Profaci, and he explains a lot of the history of what has been going on. He tells me how Maranzano and Gagliano have put up $150,000 each for the war against Joe the Boss. Besides that he says we got $5,000 a week coming in from Steve Magaddino in Buffalo and $5,000 from Joe Aiello in Chicago and I forget what else. Then one day he comes in with a sad face and says, "We're out that money from Chicago. Capone got Aiello."

(An entry in the Chicago police files dated October 23, 1930, states that Joseph Aiello, age thirty-nine, Italian, married, gang leader, was riddled with machine-gun bullets in front of 205 Kolmar Avenue when he left the home of Pasquale Prestigiocomo, alias Presto, to enter a cab. The fire was opened up on him from a "machine-gun nest" in a flat across the street, 202 Kolmar Avenue, and when he attempted to escape to the rear of the Presto home, he was felled by fire from a second nest in a window at 4518 West End Avenue.)

We're in this apartment for I'd say a month, and there's no Ferrigno. I'm beginning to wonder where the hell he is, but they explain that this is only one of the addresses he has and we got to wait. Sure enough, maybe a week later, I am sitting in the apartment with Buster from Chicago. These two guys rush in and say that Ferrigno is out in front of the building sitting on a bench in sort of a little park that runs along the parkway. They say this is the time, and I've got to drive the car. So we all go and get in it. But I don't like the setup. We got to pass right in front of the doorman, and he knows me on account of I'm always in and out of the building. "Okay," I tell Buster, "I'll drive the car, but if you see the doorman wave at me, take your gun down. If you don't, I will jerk the car. I ain't going to jail for this thing."

"What's the matter with you?" Buster says. "Are you out of your mind?"

"I don't care what the penalty is," I say. "That's the way I feel."

Buster, who was a real sharpshooter, was using a shotgun this time. He was just taking aim when the doorman waved at me and I waved back. Well, Buster was a nice boy, and he put the gun down. Now I could be in a lot of trouble. But when Buster went to see the old man—meaning Maranzano—and told him what happened, I got credit for what I done. Buster comes back and tells everyone that the old man backs me up and not to take any chances.

It was only a few days later that I was out with Buster somewhere and he left me off on the corner of Pelham Parkway in front of the apartment building. Buster left, and another car pulled up in front of me. Now we had all gotten pictures of Masseria to recognize him in case we ever see him. So, to my amazement, I saw Masseria get out of the car. I recognized him fast. This Ferrigno is with him, and they look me over suspicious-like. You got to understand this is a Jewish neighborhood, and they can see I ain't no Jew. I go into the building, and they're right behind me. In the court I got to turn right to get to my apartment, and I know they got to go left to get to Ferrigno's apartment. But they followed me into my entrance. I got into the elevator, and they got in with me. So I asked them where they want to go. They said, "Punch yours."

As I'm on the second floor, I punch six to throw them off. Going up, we were facing one another. I had my back to one wall, and they had their backs to the wall on the other side. Nobody said anything. When we got up to six, I come out of the elevator, walking like I didn't have a care in the world. As soon as the elevator door shut with them still in it, I flew down to the second floor yelling, "Joe the Boss! Joe the Boss! I just seen him."

They said it was impossible, that I was seeing things. With that, as we were talking, the one that was looking out of the window sees Masseria crossing from one side of the court to the other. He yells, "Jesus Christ, he's right! He's with Ferrigno. They're going into the apartment."

That night I was sorry I told them that I saw Masseria. They were all set to shoot him from my apartment. They had told me this wasn't to be. I got my stuff all over the joint, pictures and everything. The plan was to get him outside, not from the apartment. Now I was telling Buster to pretend not to see him. I don't know what all I was telling him. He said, "It is so important, Joe. Look, if he don't come out tonight, we'll get another apartment tomorrow."

Well, I was so worried that they gave me an assignment to watch the elevator. I was hoping and praying he don't come out that night. And he don't. Instead, a lot more of his people—maybe twenty—kept going in. They were having some kind of big meeting. Gee, was I happy.

The next day, at Valachi's behest, an apartment on the ground floor was rented. The plan now was to wait for Masseria to appear in the court, and as he passed in front of the windows of the new apartment, he was to be mowed down by a shotgun barrage. In midafternoon the meeting in Ferrigno's apartment broke up, the participants exiting in pairs. But after most of them had departed, Masseria still had not shown his face; later it was learned that Masseria had waited to be the last one out. Then when the original target, Ferrigno, came into the courtyard in the company of another Masseria lieutenant, the hidden killers decided that it was an opportunity they could not forgo. According to Valachi, the actual shooting was done by Buster from Chicago, Girolamo (Bobby Doyle) Santucci, and Nick (Nick the Thief) Capuzzi. All three fired twelve-gauge shotguns. Everyone then scattered. Valachi recalls that afterward Buster was stopped by a police officer about a block from the scene. Buster simply told him that a shooting had taken place down the street. The policeman took off, and so did Buster—the other way.

(New York Police Department records note that on November 5, 1930, at 2:45 P.M. Steven Ferrigno, also known as Samuel Ferraro, and Alfred Mineo, while leaving Ferrigno's apartment at 759 Pelham Parkway South, on the east walk of a courtyard toward the street, were fatally shot by unknown persons occupying C-1 of 760 Pelham Parkway South. The shots were fired from the ground-floor apartment through closed windows. Three shotguns were recovered.)

The day after the killing Valachi decided that the time was ripe for a long-delayed visit to his brother Anthony at Dannemora Prison in northern New York. Valachi wanted to take May along since he had seen little of her during the past several weeks. She was bedded down with a bad cold, however, and he was forced to make the trip alone. The brief time he spent with his brother was not very pleasant. "Already you could see he wasn't right in the head," Valachi says. "I didn't tell him anything I was doing. He wouldn't have understood it anyway. All he talked about was how I was to blame for all his troubles. I don't know why. We never did nothing much together. Naturally I felt bad, but what could I do? The only thing is to head back to New York."

Returning somewhat gingerly to the Pelham Parkway building, he learned from the doorman that "something" had happened during his absence. When Valachi innocently inquired what it was, the doorman related the killing of Ferrigno and Mineo in great detail and reported that the police had been all over the building. "I told them," he said, "that you were a new tenant who was away on a trip. They told me to tell you when you came back to go over to the precinct house. You know, Mr. Valachi, it's terrible. A lot of tenants have moved out because of this."

"Yeah," Valachi said, "I can imagine. I'm thinking of moving myself. A thing like this gives the place a bad name. Well, don't worry. I'll go see the cops, but there ain't nothing I can tell them."

Valachi immediately called a moving company and arranged to have all his belongings put into storage. Then he went strolling along Lexington Avenue in Harlem looking for friends. None seemed to be in sight. Suddenly an automobile screeched to a stop next to him. At the wheel was a Gagliano man, Frank (Chick 99) Callace, who frantically motioned for him to get in the car. "Jesus Christ," Callace said, "are you crazy walking around like that? We are marked. Joe the Boss is on to us after that business on Pelham Parkway. I don't know how you made it so far."

"Oh, my God," Valachi replied. "Me walking up and down like that. I didn't know. I was up in Dannemora seeing my brother."

"Well," Callace said, "light some candles. You are lucky."

Callace then drove Valachi to an apartment in the Bronx, where he found two members of his 1929 burglary gang, Nick Padovano and Salvatore (Sally Shields) Shillitani, both of whom he had recommended to Gagliano during the recuitment drive to fight Masseria. "Stay put," Callace said. "You're going upstate to meet the rest of us and the top guy. I'll let you know when."

The moment of Valachi's initiation into the Cosa Nostra had arrived. Two days later Callace came for them:

I was so worried waiting for him I didn't even take a chance calling May to see how she was.

Chick said, "Get ready. We're going on a ninety-mile trip." He knew the way and did all the driving. Besides me, there was also Sally and Nicky. We were a little nervous and didn't do much talking. We had an idea of what was going to happen.

I never knew where we were when we got there, but the house was what you would call Colonial style. Anyway, it had two stories and was painted white, and it was in the country. I don't know whose house it was. What I don't know, I don't know. It was night, and I couldn't make out any other houses nearby. I remember when we went in, Chick took us into a little room on the right. The "Doc"—that's all I know him by; he was with us for a while at the Pelham Parkway apartment—and Buster from Chicago came right in and started bullshitting with us for a minute about this and that. Then me, Sally, and Nicky were left alone. After a time some guy, I forget who, comes to the door. He waves at me and says, "Joe, let's go."

I follow him into this other room, which was very big. All the furniture was taken out of it except for a table running down the middle of it with chairs all around. The table was about five feet wide and maybe thirty feet long. Now whether it was one table or a lot of tables pushed together, I couldn't tell, because it was covered with white cloth. It was set up for dinner with plates and glasses and everything.

I'd say about forty guys were sitting at the table, and everybody gets up when I come in. The Castellammarese and those with Tom Gagliano were all mixed up, so they are one. I don't remember everybody. There was Tommy Brown—you know, Tommy Lucchese, I never heard anyone call him "Three-finger Brown" to his face. There was also Joe Profaci and Joe Bonanno and Joe Palisades—real name Rosato—and Nick Capuzzi and Bobby Doyle and The Gap and Steve Runnelli and others too numerous to mention.

I was led to the other end of the table past them, and the other guy with me said, "Joe, meet Don Salvatore Maranzano. He is going to be the boss for all of us throughout the whole trouble we are having." This was the first time I ever saw him. Gee, he looked just like a banker. You'd never guess in a million years that he was a racketeer.

Now Mr. Maranzano said to everybody around the table, "This is Joe Cago," which I must explain is what most of the guys know me by.*

---

*As a boy, Valachi was noted in his neighborhood for his ability to build makeshift scooters out of wooden crates. This earned him the nickname Joe Cargo, which later in his criminal career was corrupted to Cago.

Then he tells me to sit down in an empty chair on his right. When I sit down, so does the whole table. Someone put a gun and a knife on the table in front of me. I remember the gun was a .38, and the knife was what you would call a dagger. After that, Maranzano motions us up again, and we all hold hands and he says some words in Italian. Then we sit down, and he turns to me, still in Italian, and talks about the gun and the knife. "This represents that you live by the gun and the knife," he says, "and you die by the gun and the knife." Next he asked me, "Which finger do you shoot with?"

I said, "This one," and I hold up my right forefinger.

I was still wondering what he meant by this when he told me to make a cup out of my hands. Then he put a piece of paper in them and lit it with a match and told me to say after him, as I was moving the paper back and forth, "This is the way I will burn if I betray the secret of this Cosa Nostra." All of this was in Italian. In English Cosa Nostra would mean "this thing of ours." It comes before everything—our blood family, our religion, our country.

After that Mr. Maranzano says, "This being a time of war, I am going to make it short. Here are the two most important things you have to remember. Drill them into your head. The first is that to betray the secret of Cosa Nostra means death without trial. Second, to violate any member's wife means death without trial. Look at them, admire them, and *behave* with them."

I found out later that this was because sometimes in the old days if a boss fell for a soldier's wife, he would have the poor husband killed, whether she liked it or not. Now I was told that this wasn't an everyday thing, but once is enough. Right?

Mr. Maranzano then says, "Everybody up. Throw a finger from zero to five." So all the guys around the table threw out their right hand at the same time. Some of them had no fingers out; some had two or three, the limit, naturally, being five. When all the fingers are out, he starts adding them up. I forget what it was. Let's say they came to forty-eight. So Mr. Maranzano starts with the first man on his left and keeps counting around the table, and when he got to forty-eight, it fell on Joe Bonanno, also known as Joe Bananas. When Mr. Maranzano saw where the number fell, he started to laugh and said to me, "Well, Joe, that's your *gombah*"—meaning he was kind of my godfather and was responsible for me.

So Joe Bananas laughs too, and comes to me and says, "Give me that finger you shoot with." I hand him the finger, and he pricks the end of it with a pin and squeezes until the blood comes out.

When that happens, Mr. Maranzano says, "This blood means that we are now one Family." In other words, we are all tied up. Then he explains to me how one member would be able to recognize another. If I am with a friend who is a member and I meet another friend who is a member, but the two of them don't know each other, I will say, "Hello, Jim, meet John. He is a friend of ours." But like if the third guy is just a friend and not a member, I would say, "Jim, meet John. He is a friend of mine."

Now the ceremony is over, and everybody is smiling. I'd say it took about ten minutes. So I move away and leave the chair for the next man, who was Nicky, and there is the same routine. After him it was Sally's turn.

Then they take the knife and gun from the table, and Mr. Maranzano orders the food to be brought in. I didn't see no women, and I didn't go into the kitchen to look. I figured it was no time to be nosing around. The men around the table who were members brought the food in on big platters. First came the spaghetti *aglio ed olio*. Next there was chicken, different kinds of meat, I think also veal. There was a lot of whiskey and bottles of wine in those straw baskets, but nobody was drinking much. During the meal Sally, Nicky, and me, being the new members, talked among ourselves—mostly about how great it was to be in the mob and how we were going to put all our hearts into the "trouble."

After the coffee Mr. Maranzano got up and said, "We're here together because Joe Masseria has sentenced all the Castellammarese to death. At the same time you other guys started at your end because he had Tommy Reina killed without justice. So now we are all one. We're only a few here, but in a month we'll be four or five hundred. We have to work hard. The odds are against us. The other side has a lot of money, but while they're enjoying themselves, we'll eat bread and onions. You all will be placed in different apartments around the city. We will have spotters out on the street. These spotters will have the telephone number of our main headquarters. Headquarters will have each of your numbers. When a call comes in from the Bronx, for instance, that somebody has been spotted, the apartment we have in the Bronx will get a call. And when that call comes, you have to respond as fast as you can. Each of you new members will be placed with someone who knows what the enemy looks like. Of course, you have been given a picture of Masseria. He's the most important one. I also want to tell you that the business at Pelham Parkway has got them confused. They don't know how we found out about the meet they were having, and that's in our favor. Already they don't know who to trust. We must concentrate on getting their main bosses, and we must get Masseria himself. There will be no deal made with Joe Masseria. The war will go on for ten years if we don't get him."

After the meeting was over, The Gap took me and Sally to the apartment which was picked out and furnished for us on the Bronx River Parkway.

Throughout the rest of 1930 and into 1931, the Castellam-marese War raged on, and before the bloodletting was over, some sixty bodies would litter U.S. streets. Although nationwide in scope, the crucial battle was waged in New York City between the Maranzano and Masseria forces. Valachi's first call to action came, as he puts it, in the "wee hours"—around five o'clock one morning—when a spotter reported by telephone to the apartment that two Masseria men, Joseph Rao and Big Dick Amato, together with two girls, were at the Pompei Restaurant in Harlem.

Valachi could not have been more delighted. Rao, he had learned, had been the one behind the false charge that Valachi was the driver of the car for the Irish gang during the shooting incident years before that subsequently caused him so much grief. Valachi and Shillitani immediately drove to the restaurant. They arrived just as Rao was getting into a car with his lady friend. Valachi pulled alongside. "Take him," Valachi snapped, but Shillitani, after aiming his shotgun, "froze in his tracks." He had neglected to release the safety catch on his shotgun. Furious, Valachi yanked out a pistol and fired four shots at Rao, who by now was scampering down the street. "I'm so mad," Valachi recalls, "that I can't aim straight. I got him once, in the ass, but it wasn't enough to stop him." Then, with people pouring out of the restaurant, Valachi was forced to drive off hurriedly, still cursing the hapless Shillitani.

Even in the company of a more experienced hand named Steve Runnelli, Valachi's luck did not improve:

I was taken out of the apartment on the Bronx River Parkway, and they put me right in Mr. Maranzano's headquarters, which is in Yonkers. We are told that we are supposed to go after the top people with Masseria and not to fool around with the little guys. That is the way to get the war over fast.

So now one day Steve Runnelli and I were riding up in Harlem when Steve spots a man in a Lincoln coming the other way. He tells me to swing around fast. I said, "Who is that?" and he said, "He's a big boss with the others." "Oh," I said and swung the car around and pulled up to the Lincoln. When I was right alongside the Lincoln, Steve fired one shot at him, and I saw this man—I found out later that his name was Paul Gambino—go with the shot. I tried to tell Steve that I don't think we got him, but he was panicky after the first shot, and he started yelling, "Step on the gas! Step on the gas!" I was disappointed. These mob guys were always talking big, and then, when the action comes, they fold up. As I pulled away, my rear bumper got caught in his front bumper. So I was pulling and pulling for about half a block before I got loose, with Steve yelling all the time, "Get the hell out of here!"

I let Steve off someplace on Southern Boulevard, and I went back to Yonkers where Mr. Maranzano was. He was waiting for me. Boy, was he mad! He seemed to know already what happened, and he said, "What did you go after that guy for? What's the matter with you?"

I said, "Steve told me he was a big boss."

Mr. Maranzano started cursing. He said, "This guy is a nobody." I told him not to worry, I didn't think the guy was hit. With that, as I am telling him this, the phone rang. After he took the call, he said to me, "You are right. The bullet only clipped his ear. He got out of that Lincoln, I just heard, and walked away."

Well, everything turned out for the best. Paul Gambino wasn't nothing, but his brother Carlo was a big man with the Masseria group. When Carlo got word of the shooting, figuring he might be next, he brought all his people over with us.

Not all the contracts Valachi was involved in had such Keystone comedy aspects. Maranzano was particularly adamant about killing a Masseria lieutenant, Joseph (Joe Baker) Catania. Valachi was first told about it by Buster from Chicago. "You know this Joe Baker?" Buster asked.

"Yeah," Valachi replied. "I know him from the old neighborhood. I like him."

"Don't let the old man hear you say that," Buster advised Valachi. "He has a personal grudge against him. This Baker hijacked some of the old man's alcohol trucks, and now he has to go."

Consoling himself with the thought that Baker was the nephew of a bitter enemy of his own, Ciro Terranova, Valachi said, "Well, if he's got to go, what do I care?" On a January morning shortly after this exchange, Valachi was picked up by Buster and, along with two other gunmen, was taken to an empty apartment in the Fordham section of the Bronx. The apartment was diagonally across the street from a bail bondsman's office, which Baker had apparently been using as a collection point for the money from his various rackets. He had been seen entering the office every morning for several days, staying five or ten minutes, and then leaving with a brown paper bag in his hand. "Today," Buster told Valachi, "we're just going to check the information out." At about 10 A.M. a man rounded the corner, walked the eight feet or so to the entrance door, and went inside. "That's him," Valachi said to Buster. "That's Joe Baker, all right." About ten minutes later, as predicted, Baker emerged from the office carrying a paper bag and disappeared around the corner.

Throughout the rest of the week Baker continued his routine. His waiting killers, however, faced several problems. Baker always approached the office from the same direction, and the eight feet he walked after turning the corner did not give Buster time enough to shoot. The same situation prevailed when he exited. Worst of all, the apartment they were in was on the fifth floor, and Buster, with only seconds at his disposal, was afraid that the range was too great for an effective shot.

Valachi thought he had an answer to the problem. He had noticed a ground-floor apartment in the same building that seemed to be empty, and he suggested that they force their way into it the next morning before Baker arrived. Buster agreed at once. But when they did, the killers were momentarily dumbfounded. Three house painters were hard at work. While Buster headed for the window, Valachi and the other two men lined up the painters against the wall at gunpoint. Suddenly he heard Buster call to him, "Come here, Joe. Who's that woman with him?" Valachi rushed to the window just in time to see Baker's companion and said, "Jeez, that's his wife." Then he left the apartment to make sure that their car, parked around the block, was ready to go. A few minutes later he heard shots, and as soon as his pals arrived, they sped away. "How did it go?" Valachi asked.

"He came out of the office with his wife," Buster replied. "He kissed her in front of the office, and I was worried I wouldn't get a shot. But he turned and went for the corner. She was just standing there watching when I got him. I don't think I missed once. You could see the dust coming off his coat when the bullets hit."

"It's too bad the wife had to see him go," Valachi said.

"Yes, but would we have got another chance with those painters spreading the word around the neighborhood?"

Wife or not, Maranzano could not have been more delighted when the news came that afternoon that Baker died in a Bronx hospital. Later a Maranzano spy reported that Terranova had placed his hand on his nephew's coffin at the funeral parlor and swore he would avenge his death. Maranzano, enraged at this, sent the spy back to look into the possibility of cornering Terranova at the wake, but, as Valachi sadly recalls, "It was impossible to do anything there."

(Case No. 122 of the 46th Squad, New York Police Department, states that on February 3, 1931, at 11:45 A.M. one Joseph Catania, alias Joe Baker, of 2319 Belmont Avenue, Bronx, male, white, twenty-nine years old, while walking in front of 647 Crescent Avenue, was shot six times in the head and body, causing his death in Fordham Hospital, to which he was removed. Before he died, in spite of his mortal wounds, he could not or would not identify the perpetrators of this crime.)*

In the weeks following the Baker killing, according to Valachi, the tide of battle swelled in Maranzano's favor. There had been so many defections to him, like that of Carlo Gambino, that his forces actually outnumbered the enemy. Moreover, for those who still sided with Masseria there was an increasing economic problem because of the struggle, their well-fixed rackets were rapidly becoming a shambles. It was then that Valachi first heard that there was "talk of peace in the air." But Maranzano would enter into no negotiations as long as Masseria remained alive. Finally two of Masseria's most trusted sidekicks, Charley Lucky Luciano and Vito Genovese, secretly turned against him. In return for their promise to have Masseria killed, Maranzano agreed to halt the war.

---

*Catania's funeral was the most lavish held in the Bronx up to that time—and probably since. His coffin, solid bronze, cost $15,000. It was, according to one report, the "finest casket ever given a gunman in New York." Five funeral directors shared the honors. More than 10,000 people turned out to watch the cortege. Forty automobiles were required to carry the flowers alone. The most impressive floral decoration came from Ciro Terranova and other close associates: it was thirteen feet high, made up of red and white roses, and featured the inscription "Our Pal."

Masseria literally never knew what hit him. He was invited by Luciano to dine one afternoon in a Coney Island restaurant called Scarpato's. From all accounts Joe the Boss, surrounded by trusted aides, had a fine time during the meal—the last one he ever ate.

(Case No. 133 of the 60th Squad, New York Police Department, notes that at about 3:30 P.M. on April 15, 1931, Giuseppe Masseria, also known as Joe the Boss, last known residence, 65 Second Avenue, New York City, while sitting in a restaurant at 2715 West 15th Street, in the Coney Island section of Brooklyn, was shot in the back and head by unknown persons who escaped.)

Luciano was still at the restaurant when the police arrived. Luciano explained that he was unable to be of much help. After the meal, he said, he had suggested a game of cards. He and Masseria played for approximately forty-five minutes while the restaurant emptied. Then Luciano excused himself for a trip to the men's room. As he was washing his hands, he heard some "noise," hurried out to see what was going on, and found Masseria slumped over the table.

None of the restaurant staff could identify any of the others in attendance at Masseria's execution. But besides Luciano and Vito Genovese, Valachi says, they included his old friend Frank Livorsi and Joseph Stracci, alias Joe Stretch.

Thus Salavatore Maranzano, always a shadowy figure as far as the police were concerned, became the undisputed chieftain of the Italian underworld in America. His ascension to power also marked the organization of the modern Cosa Nostra:

Mr. Maranzano called a meeting. I was just notified. I don't remember how, but I was notified. It was held in the Bronx in a big hall around

Washington Avenue. The place was packed. There was at least four or five hundred of us jammed in. There were members there I never saw before. I only knew the ones that I was affiliated with during the war. Now there were so many people, so many faces, that I didn't know where they came from.

We were all standing. There wasn't any room to sit. Religious pictures had been put up on the walls, and there was a crucifix over the platform at one end of the hall where Mr. Maranzano was sitting. He had done this so that if outsiders wondered what the meeting was about, they would think we belonged to some kind of a holy society. He was just hanging around, waiting to speak, while the members were still coming in.

Joe Profaci had given me Mr. Maranzano's pedigree. He was born in the village of Castellammare and had come over here right after the First World War. He was an educated man. He had studied for the priesthood in the old country, and I understand he spoke seven languages. I didn't know until later that he was a nut about Julius Caesar and even had a room in the house full of nothing but books about him. That's where he got the idea for the new organization.

Mr. Maranzano started off the meeting by explaining how Joe the Boss was always shaking down members, right and left. He told how he had sentenced all the Castellammarese to death without cause, and he mentioned the names of a half a dozen other members and bosses who had suffered the same thing.

Well, some of these names I didn't know or never even heard of, but everybody gave him a big hand. He was speaking in Italian, and he said, "Now it's going to be different." In the new setup he was going to be the *Capo di tutti Capi*, meaning the "Boss of all Bosses." He said that from here on we were going to be divided up into new Families. Each Family would have a boss and an underboss. Beneath them there would also be lieutenants, or *caporegimes*. To us regular members, which were soldiers, he said, "You will each be assigned to a lieutenant. When you learn who he is, you will meet all the other men in your crew."

Then he tells us how we are going to operate, like if a soldier has the need to see his boss, he has to go first to his lieutenant. If it is important enough, the lieutenant will arrange the appointment. In other words, a soldier ain't allowed to go running all the time to his boss. The idea is to keep everything businesslike and in line.

Next he goes over the other rules. The organization, this Cosa Nostra, comes first, above everything no matter what. Of course we already know that death is the penalty for talking about the Cosa Nostra or violating another member's wife, but he goes over it again anyway. He tells us now that death is the penalty for telling wives anything about the outfit and also that an order going from a boss to a lieutenant to a soldier must be obeyed or you die.

Now there are other rules where death ain't the penalty. Instead, you are "on the carpet"—meaning you have done wrong and there is a hearing to decide your case. The most important one is that you can't put your hands in anger on another member. This is to keep one thing from leading to another. Remember, we are just getting over our trouble, and that's what Mr. Maranzano talked about next.

"Whatever happened in the past is over," he says. "There is to be no more ill feeling among us. If you lost someone in this past war of ours, you must forgive and forget. If your own brother was killed, don't try to find out who did it to get even. If you do, you pay with your life."

The table of organization set forth by Maranzano was subsequently adopted by Cosa Nostra Families across the country. And out of this meeting came the five-Family structure in New York

City, which still exists. The five bosses initially named by Maranzano to head them were Luciano, Tom Gagliano, Joseph Profaci, Joseph Bonanno, and Vincent Mangano. Valachi can recall only three of the underbosses selected at that time—Vito Genovese in the Luciano Family, Albert Anastasia in the Mangano Family, and Thomas Lucchese in the Gagliano Family. Valachi himself elected to join Maranzano's personal palace guard even though he had entered the Cosa Nostra under the auspices of what was now the Gagliano Family.

The switch, he says, came on impulse. At the meeting Maranzano had announced, "As for those members who have been with me, there is going to be a split. Some of the group will go back to Tom Gagliano, and some will remain with me. If there is anybody who wants to remain with me, whether he was with me before or whether he was not, as long as he was with me during the war, he is entitled to come with me now if he wants to. The ones who want to come with me, just raise your hands."

Annoyed because Gagliano had made no effort to recruit him privately, Valachi, as he puts it, "unconsciously" raised his hand. Then he saw two other Gagliano men, Bobby Doyle and Steve Runnelli, also raise their hands. Almost at once he regretted his move because of his unhappy experience with Runnelli in the Gambino shooting. But when Thomas Lucchese immediately tried to persuade him to change his mind and remain with Gagliano, he felt that to do so now entailed too much loss of face. "Why did you do what you did?" Lucchese demanded.

"Well, you didn't tip me off. I figured I wasn't wanted."

"Let's go see the old man and tell him you made a mistake."

"No," Valachi replied. "It's too late. I won't do anything. I would be ashamed of myself."

On the plus side, however, he could comfort himself with the thought that he would be with Buster from Chicago and that he was at least rid of Salvatore Shillitani, who had decided to remain with Gagliano. Whatever other doubts Valachi might have had disappeared in the excitement of a huge banquet held in Brooklyn to honor Maranzano—in spirit and in cash. As a sign of their obeisance Cosa Nostra bosses throughout the United States purchased tickets to the affair. Even Al Capone sent $6,000. In all, according to Valachi, about $115,000 was collected. As representatives of various Families arrived at the banquet, each threw his contribution on a table. "I never saw such a pile of money in my life," Valachi recalls. Afterward he began rotating on duty with Maranzano as a chauffeur and bodyguard. But tranquillity in Maranzano's reign was short-lived:

Mr. Maranzano's legitimate front was a real estate company. The offices were in the Grand Central Building at 46th Street and Park Avenue. Around six months after Joe the Boss got his—that makes it around the first part of September—he ordered us not to go into the offices carrying guns, as he had been warned to expect a police raid.

I didn't like that order. One of the other fellows, I think it was Buster, said, "What do you mean?"

"I don't know," I said. "I just don't like the idea. If something happens, we are helpless."

He said, "Talk to the old man about it."

Well, I was biding my time to do it, but I was waiting for the right moment. Mr. Maranzano wasn't somebody you just started telling what to do. Anyway, a couple of days later I was in the office, and he told me to come by his house in Brooklyn that night. It was on Avenue J; I don't remember the number.

I got there about nine o'clock. When I went into the living room, he was bent over bandaging a cut on the foot of his youngest son. I'd say the kid was around eight. Mr. Maranzano got right to the point. "Joe," he said, "I hear you're wondering why you didn't get a bigger piece of the take from the banquet." He was right. All I was getting was my expenses and maybe $100 a week. To tell the truth, I was doing a little burglarizing on the side.

He goes on to say, "Don't worry. You'll get your share, and more. But we are holding onto the money right now because we have to go on the mattress again." In other words, he is telling me we have to go back to war. You see, during the Castellammarese trouble we had to take mattresses with us as we were moving from one apartment to another. Sometimes we only had a minute's notice, and so you needed a mattress to sleep on. That is our meaning of going on the mattress.

I'm still listening as he explains why. He said, "I can't get along with those two guys"—he was talking about Charley Lucky and Vito Genovese—"and we got to get rid of them before we can control anything." He talked about some others who had to go too—like Al Capone, Frank Costello, Willie Moretti from Fort Lee, New Jersey, Joe Adonis, and Charley Lucky's friend from outside the Cosa Nostra, Dutch Schultz.

By controlling things, he meant the Italian lottery, which was very big then, the building unions, bootlegging, bookmaking, all that kind of stuff. Dutch Schultz, who was a Jew boss, had the biggest numbers bank in New York, and Charley Lucky ran the downtown lottery.

Gee, I wanted to say, who wants to control everything? You got to remember it's just a few months since we are at peace. All I wanted was to make a good living. But naturally, I dared not say anything.

Then Mr. Maranzano tells me that he is having one last meeting the next day at two o'clock with Charley Lucky and Vito. This is my chance to bring up the thing about no guns. "Gee," I said, "this is no time to be putting yourself in jeopardy. Suppose those guys know something's in the wind?" Well, he's talking so much about what we are going to do and how big we are going to be, that he don't pay much attention, but finally he tells me, "All right, call the office at a quarter to two to see if I need you."

I went home and spent the night tossing and turning. I got all kinds of reasons to worry. If something happens to Mr. Maranzano, I'm finished, too. I ain't happy, but I got to go along. All I can do is wait to check in. That afternoon I called the office and this guy Charlie Buffalo, who is one of the members with us, answered the phone and said that everything was fine and that I don't have to go down there. Right after that The Gap comes by—he had stayed with Gagliano, which is another reason why I should have—and says, "Hey, I've been looking for you. I got a couple of new girls over in Brooklyn. Let's go over and spend some time with them."

I said, "Good idea. I have nothing to do." So we went over to Brooklyn and fooled around with the girls until about midnight, and the four of us decided to come back to Manhattan to eat. We went to this restaurant Charley Jones had on Third Avenue and 14th Street. When we are in the restaurant, I notice something funny is going on. First one guy, and then another, is walking in and looking us over. I said to The Gap, "Do you see what I see?" He says, "Yes, I don't know what it's about." Now I go over to Charley Jones. He is no mob guy, but he has connections and knows a lot. He whispers, "Joe, go home."

That's all I have to hear. I said to The Gap, "What do you think?" and he says, "We'll break it up." So we gave the girls money to get back to Brooklyn. The Gap stayed with Charley, and I drove home alone. On the way up Lexington Avenue I stopped for a newspaper. I just put it down beside me on the seat. I was driving very slow and thinking. I just could not figure out what was going on. It's hard to explain—I was worried, and I did not know what I was worried about. When I got home, I sent my kid brother out to put the car in the garage. Then I sat down in a chair and opened up the paper, and there it was. All about how the old man had been killed in his Park Avenue office that afternoon.

The story had said that some men pretending to be detectives walked into Mr. Maranzano's outer office—it was kind of a waiting room—and lined up everyone who was there against the wall. Then two of the fake detectives went inside, where they shot him and cut his throat. The first thing I remembered was Mr. Maranzano telling us not to carry guns in the office as there might be a police raid. Then I read that when the real bulls came, they found Bobby Doyle kneeling by the old man and had pulled him in as a material witness.

Now I tell myself no wonder. The Gap took me to Brooklyn. He must have known all about it and kept me out of the neighborhood. You can imagine how I felt.

5

The Murder of Maranzano was part of an intricate, painstakingly executed mass extermination engineered by the same dapper, soft-spoken, cold-eyed Charley Lucky Luciano who had so neatly arranged the removal of Masseria just months before. On the day Maranzano died, some forty Cosa Nostra leaders allied with him were slain across the country, practically all of them Italian-born old-timers eliminated by a younger generation making its bid for power.

Valachi immediately went into hiding when he learned that three men he had recruited for Maranzano—Stephen (Buck Jones) Casertano, Peter (Petey Muggins) Mione, and John (Johnny D) DeBellis—had narrowly escaped being gunned down while walking along Lexington Avenue. "If they want them," Valachi recalls, "I got to figure they want me, too." He first turned to Nick

Padovano, with whom he had been initiated into the Cosa Nostra. A reluctant Padovano let him stay overnight but tearfully begged him to depart in the morning. "I'm supposed to let them know if I see you," Padovano said. "So for Christ's sake, don't tell anybody you was here."

In desperation Valachi called the son of Gaetano Reina and reminded him how he had fought to avenge his father's death at the hands of Masseria. Young Reina, himself a member of the Gagliano Family, recognized the debt and agreed to sequester Valachi in the attic of his house. Once there, Valachi, again gambling on his past service, called Thomas Lucchese. "Stay put," Lucchese told him. "You're safe where you are. I'll see what I can do."

Two days later Lucchese called back and said that he and Gagliano wanted to see him. When he arrived, he recalls that Lucchese did most of the talking. Lucchese began by asking Valachi if he knew that Maranzano had been hijacking liquor trucks belonging to Luciano. "Tommy," Valachi replied, "I don't know nothing, so help me God."

Lucchese continued to press him. Had he seen large amounts of cash being split up in Maranzano's office? "Yes," Valachi said, "but I never asked any questions. You know how it was with the old man." Was he aware of the fact that Maranzano had hired a free-lance killer outside the Cosa Nostra, Vincent (Mad Dog) Coll, to rub out Luciano and Vito Genovese? "No," Valachi insisted, "I didn't know, believe me! I never heard any of these things. What's it all about?"

"Joe," Valachi quotes Lucchese as saying, "I think you are in the clear, but I had to judge you to your face. I must tell you that the old man went crazy and was going to start another war. He couldn't leave well enough alone." On this point Valachi remained understandably discreet as Lucchese continued, "Now Tom Gagliano wants you to come back with us, providing you are telling the truth. But we are only interested in you, not Bobby Doyle or those three guys you brought to the old man. Think it over for a few days while we consider your story. Until then you got nothing to worry about."

At least able to vacate the Reina attic, Valachi returned to the East Harlem apartment he maintained for his mother, younger brother, and sisters. But almost at once there were new complications when Buster from Chicago, who had vanished after the Maranzano killing, paid him a visit late one night. "What are we going to do?" Buster said. "Are we going to fight? If we don't, they'll just take us anyway, one by one."

"I don't know what to think," Valachi said. "Maybe you're right. Let's wait for Bobby Doyle. Stay out of sight, and don't get into any arguments." Buster apparently disregarded the advice. Within a week, according to Valachi, he was killed on the orders of Luciano and Genovese on Manhattan's Lower East Side. The disposal of his body remains a mystery; it is the one such gangland slaying cited by Valachi for which there are no police records. Meanwhile, Doyle was finally released from custody after the Maranzano murder:

All this time I am waiting for Bobby to come out, and now I find out what happened in the Park Avenue office. He explained to me that they were all sitting around in the outer office when these four Jews walk in and flash badges. I think one was with Dutch Schultz, but the others were really Meyer Lansky's boys. Charley Lucky could use them because his Family had sided with Lansky in his Jew war with Waxy

Gordon. Anyway, Bobby said that what with all the yelling, the old man stuck his head out of the inner office to see what was going on, and one of the fake bulls says, "Who can we talk to?" and the old man says, "You can talk to me." So two of them go in with him, and the other two keep an eye on the crowd.

A long time after this I met one of the boys who went in with the old man. I met him at a racetrack. His name was Red Levine. I said, "I heard you were up there." Levine said, "Yes, I was there. He was tough." He told me the idea was to use a knife, so there wasn't any noise, but the old man started fighting back, and they had to use a gun first.

Naturally Bobby Doyle don't know nothing about this as he was in the waiting room. All he knows is he hears a couple of shots, and this Levine and the other Jew boy come running out and tell everybody to beat it. But Bobby says he ran in to see if the old man still had a chance. That's how come the cops picked him up.

(New York City police records show that at 2:50 P.M. on September 10, 1931, one Salvatore Maranzano, male, white, of 2706 Avenue J, Brooklyn, was shot and stabbed to death in the office of the Eagle Building Corporation, rooms 925 and 926, at 230 Park Avenue. Perpetrators were four unknown men posing as police officers. Cause of death: four gunshot wounds and six stab wounds.)

Now I say to Bobby, "Buster wants to fight. How do you feel?" He says, "No, all the guns are pointed at us. Give me a few days. I know Vito pretty good." I said, "What's on your mind?" and he says, "We may go with Vito." To tell the truth, I was shocked, but I said, "Okay, what do you want me to do?" He said, "Just call me in a couple of days."

In the meantime, I get in touch with my good friend The Gap, and we took a ride. I said to him, "Look, what will I do? I understand Mr. Maranzano was doing a lot of dirty work, and anyway it looks like we ain't going to be making a comeback. Tom Gagliano wants me to go back with him, and now Bobby is saying we may go to Vito. Give me your advice."

The Gap says, "Go with Vito."

I'm still not so sure. But The Gap explains that with me being so close to the old man, the best way to get out of the spot I'm on is to go with Vito, who is with Charley Lucky. That way, nobody can ever question where I stand.

So I call Bobby, and he says he has made a meet with Vito Genovese for tomorrow. Then he says, "You know where you can get them other guys?"—meaning Buck Jones, Petey Muggins, and Johnny D. Well, we all meet with Vito at the Cornish Arms Hotel on West 23d Street downtown. When we get there, Vito speaks to us. He says, "I want to take you boys along with me because I want to see you get the respect due you." In other words, he means he would fix it so we wouldn't have to worry about nobody going after us because we had been with Mr. Maranzano.

"By you being with us," he says, "you have prestige, and everything will be just the same." He was showing us that his people are in power *now*. After all, he is underboss to Charley Lucky. Believe me, I was thinking hard. You don't know what it's like to be on the top floor one day and in the cellar the next. I heard that at the old man's wake nobody showed up.

Then Vito says, "You know, we made it by minutes." That's when I learned what was behind it all. Remember when I went out to Brooklyn to see the old man and he told me he was going to have an appointment the next day with Charley Lucky and Vito? Well, it was to set them up. He had an Irish guy, Vincent Coll, coming in to get them. He was using Coll, as this mad dog was going around shooting up everybody in New York anyway, so who would figure Mr. Maranzano was in on it? Coll was just coming into the building when the old man got his, and there had to be a double cross somewhere. The rat smelled like Bobby Doyle, but there was nothing I could say.

It was lucky I wasn't in the office. The Gap saved my life by getting me out of the way. But a lot of others around Mr. Maranzano who got caught sleeping slept forever. They got Jimmy Marino sitting in a barber's chair in the Bronx, and they threw Sam Monaco and some other top guy in New Jersey into the Passaic River. Sam had an iron pipe hammered up his ass.

(New York City police records show that at 5:45 P.M. on September 10, 1931, the same day Maranzano was killed, James LePore, also known as Jimmy Marino, male, white, while standing in the doorway of a barbershop at 2400 Arthur Avenue in the Bronx, was shot six times in the head and body, causing his death. New Jersey police records show that on September 13, 1931, two bodies washed ashore in Newark Bay. One body was identified as Samuel Monaco; the other was Louis Russo. Both men had their heads crushed and their throats cut. They were wrapped in sash cord and weighted down with sash weights. A missing persons bulletin had been sent out re Monaco on September 10. His car was eventually discovered parked on 46th Street near Park Avenue, New York City.)

I asked Vito how come there had to be these killings, and he said that whenever a boss dies, all his faithful have to go with him, but he explained that it was all over now, and we didn't have a thing to worry about. Then he went on to explain how Maranzano was hijacking

Charley Lucky's alcohol trucks and I don't know what else. Vito said, "If the old man had his way, he would have had us all at each other's throats."

So I said, "If you people thought that he was doing so much wrong, why didn't you approach me?"

"We couldn't approach you," Vito said. "The old man pulled 100 percent with you." In other words, he was saying he feared that I would warn Maranzano, and he was right. I told him so. I said, "If I am coming on your side, I want to come not being known as a double crosser." I don't like being with Vito, but what can I do?

So now I am in Charley Lucky's Family. Vito takes us down to the Village and introduces us to Tony Bender, real name Anthony Strollo. Now you got to remember that every Family is split up into what we call crews. Vito put me in Tony's crew. This made him my lieutenant.

Salvatore Luciano earned his nickname the hard way. Once, as he was rising through racketeering ranks in the Italian underworld, he was kidnapped by rival hoodlums who were after a cache of narcotics he had stashed away. Luciano was taken to a deserted section of Staten Island and hung up by the thumbs from a tree. He was tortured with razors and lighted cigarettes but refused to talk. Believing him near death, his inquisitors left him to his fate. Luciano lived, however, and the legend of Charley Lucky was born. From such untidy incidents he would eventually become the most powerful chieftain the Cosa Nostra has ever known, holding court in an elaborate suite in the Waldorf-Astoria Hotel, where he resided as "Mr. Charles Ross." In time no significant racket in New York—bookmaking, numbers, industrial and labor extortion, narcotics, loansharking, etc.—operated without his cooperation or cut.

After the initial purge of the Maranzano "faithful," the massive bloodletting that had racked the Cosa Nostra was largely curbed. Indeed, according to Valachi, Luciano moved swiftly to reduce the special tensions that existed in the New York area by establishing *consiglieri*, or councilors, six men in all, one from each of the five New York Families, and the sixth representing nearby Newark; the function of the councilors was to shield individual soldiers from the personal vengeance of various lieutenants who might have been their targets during the Castellammarese War. The councilors, Luciano declared, had to hear the precise charge against a particular soldier before his death was authorized. If there was a tie vote, any one boss could sit in and break it. While the councilors were as often as not ignored, their formation at least projected an aura of the stability Luciano was bent on achieving.

Luciano also abolished the position of Boss of all Bosses, so dear to Maranzano's heart; the fact of his power was infinitely more important to him than its formal trappings. And the organization of his own Family symbolized, to the naked eye at any rate, the final breakdown of the old Neapolitan versus Sicilian hostility. Luciano came from Sicily; his number two man, Genovese, was born in Naples. But most important by far was the way Luciano revolutionized the scope and influence of the Cosa Nostra in the U.S. underworld. Shrewd, imaginative, and, above all, pragmatic, he abandoned the traditional clannishness of his predecessors and joined in cooperative ventures with such non-Italian criminal associates as Dutch Schultz, Louis (Lepke) Buchalter, Meyer Lansky, and Abner (Longy) Zwillman. At the same time, however, he carefully maintained the identity of the Cosa Nostra since, for Luciano, peaceful coexistence was merely a step toward total domination of organized crime.

Nonetheless, to consolidate his power base, Luciano immediately had to justify his elimination of Maranzano to Cosa Nostra kingpins elsewhere in the country. "When a boss gets hit," Valachi notes, "you got to explain to the others why it was done." And to his amazement Valachi was told by Genovese that Charley Lucky desired him to testify personally against Maranzano in Chicago, where Capone, although on the verge of being imprisoned for income tax evasion, was still a force to be reckoned with.

"Why me?" Valachi asked.

"First, because you are known to be close to the old man," Genovese said. "And second, by being one of his soldiers, you gained nothing when he went. So why should you lie?"

Valachi begged off on the ground that he wasn't eloquent enough for such a delicate chore and suggested that a more experienced member, like Bobby Doyle, be sent instead. Actually the battle-weary Valachi, ever cautious, feared that this sort of assignment held unforeseen perils should some new coup be in the making. However ill-founded his misgivings were, he made his point. Genovese agreed, and Doyle, delighted at the status he would gain from the trip, was dispatched to appear before, as Valachi describes them, "our friends in Chicago."

Soon afterward Valachi, in partnership with Doyle, received his first tangible benefit as a member of the Luciano Family. Frank Costello, then a wily Luciano lieutenant, had long since forsaken muscle for political influence to advance his career. New York City was ideal for Costello in those days. Under the dubious reign of Mayor James J. Walker, he had "opened up" the town for slot machines and ran their operation.

Valachi and Doyle approached their own lieutenant, Tony Bender, and asked if there was a chance for them to have a "few machines." Bender took them to the offices of a scrap company on Thompson Street in Greenwich Village which Vito Genovese used as his legitimate front. It happened to be a day when Luciano was present. When the door to the inner office was opened, Valachi suddenly felt himself being pushed in by Bender while Doyle, apparently suffering an acute case of cold feet, stayed outside. Luciano, in the grand manner, spoke not to Valachi, but to Bender: "What's he want?"

"He wants some machines."

As Valachi waited, nervously silent in the great man's presence, he finally heard Luciano say, "Okay, give him twenty."

This did not mean actual machines; Valachi and Doyle would have to finance that themselves. It signified that Valachi was entitled to twenty stickers supplied by Costello. In theory, while slot machines were nominally illegal, any enterprising soul could install one in the rear of a candy store, a pool hall, and so forth. In fact, if a machine did not have a Costello sticker, the color of which was periodically changed, it was immediately subject to not only mob action, but police seizure. Once, according to Valachi, a patrolman walking his beat in a Manhattan neighborhood dumped, for whatever reason, a bottle of catsup down a "protected" machine and was promptly transferred to the far reaches of Queens. "Now," Valachi reflects, "it don't take too much to figure who had him sent out there."

It was up to Valachi to place his own machines. He selected the area most familiar to him, East Harlem, and he and Doyle, as soon as they had them suitably installed, were grossing about $2,500 a week. Valachi hired the brother of his former fence, Fats West, to service the machines and pick up the money. Occasionally Valachi would play one of them himself, doubling or tripling its normal take to see if the extra amount was returned to him. Fortunately for young West's health, he turned out to be an honest man.

Valachi derived just as much pleasure, if not profit, from being "recognized as a mob guy" with connections. His first opportunity along these lines came when old Alessandro Vollero was paroled from Sing Sing after serving fourteen years for the murder of Ciro Terranova's brother.* Vollero, fearful that Terranova would seek revenge, sent an emissary begging Valachi's help. "The old guy don't know nobody now," the emissary said, "but he hears you're with Vito and them others. Can you straighten things out for him?"

Valachi remembered Vollero with fondness, promised to see what he could do, and discussed the matter with Vito Genovese. At first Genovese was reluctant to get involved in such ancient history—"When the hell did this happen, twenty years ago?"—but eventually he reported that Vollero could stop worrying. When Valachi relayed the good news, nothing less would do than for him to come to Vollero's house for dinner. "It was really something," Valachi says. "He had the whole family lined up to greet me. He called me his savior. Well, we ate, and it was the last time I saw him. I heard later he went back to Italy and died in peace."

Now twenty-six years old, newly affluent and respected, Valachi was ready to get married. His affair with his dance-hall mistress, May, had continued, but he discovered she had been unfaithful to him during his frequent absences in the Castellammarese War. "I told her that I'd stay on with her for a while," he says, "but she could forget about anything permanent."

---

*Vollero, according to Sing Sing records, was released on April 28, 1933.

True love began to bloom for him when he was hiding out in the Reina household and met the dead racketeer's eldest daughter, Mildred. Then twenty-two, she would come up to the attic each afternoon to keep him company. And once out of hiding, he became a frequent visitor in the Reina home. But Mildred's mother, brother, and uncles, as soon as they realized what was happening, did not cotton at all to the idea of Valachi as a prospective bridegroom. Romeo and Juliet had nothing on the tribulations endured by Joseph and Mildred. Before it was all resolved, their romance featured a foiled elopement, an attempted suicide, and, finally, the intervention of Vito Genovese himself:

When I was in the attic, Mildred would come up the steps through sort of a trapdoor and talk to me. She told me that she had heard a lot about me. I asked her who it was that told her, and she said Charley Scoop, who was a guy I used to sell dresses to when I was out stealing.

Then after everything was straightened out with Vito about the Maranzano business, I was invited back to the house for supper by her brother. I accepted and brought Johnny D along with me. All through supper Johnny kept kicking me under the table and whispering that Mildred is nuts about me and can't help showing it. When we left, I told Johnny that Mildred was a beautiful girl, all right, but her family looked pretty strict to me, and I better not get mixed up with them. Anyway I figured I didn't know how to act around her as I was only used to hanging around with dance-hall girls. I was going to forget the whole thing, but this idea of finally settling down began to get to me. I tried going out with some of the nice girls in the neighborhood. It didn't work. They all looked bad next to Mildred.

A couple of days later Bobby Doyle and me started talking about her somehow. Bobby knew the family pretty good. He even knew the old man before Masseria had him killed. Bobby told me that his wife, Lena, was saying that Mildred liked me a lot. "Oh," I said, "Mildred is okay, but that mother she's got is a tough woman, and I think the brother is like the mother. I don't even want to think about her uncles."

"What are you worrying about them for?" Bobby said. "It's what the girl says that counts." I said I don't know. After all, my people are from Naples, and they are Sicilian.

Well, after one thing and another, Bobby's wife tells me that Mildred wants me to speak to her brother about us. I said I wouldn't. Finally Lena says that Mildred is willing to take off with me. By that she means that Mildred was ready to leave home to marry me. At this time I was living at 335 East 108th Street. The building had just been done over, and it had steam heat and hot water, and we had four rooms, so it was a place to stay for a while. I asked Lena how I would know when Mildred was home alone, and she says leave everything to her.

When she gives me the word, I head for Mildred's house. She is all packed and ready to go, but then her sister Rose says that she has to go, too, or they'll beat her to death. After I heard this, I told Mildred to unpack. I said, "Okay, I'll talk to your brother."

I was amazed when her brother said that it was all right by him to marry her. But the next thing I hear is that Mildred's in the hospital. She had swallowed a bottle of iodine. I found out why. Her brother told her that I didn't want anything to do with her. So now I know how they work. Her brother says okay to me and tells her just the opposite. Lena told me not to go up to the house, as there was a big commotion going on there. I told her maybe I better drop the whole thing, but she says, "A fine guy you are. Here she's taken iodine for you, and you're talking this way."

Well, I went down to see Vito at his office on Thompson Street next to that junkyard he owned and explained the whole mess to him. He told me he would pass the word around that he wants to see Mildred's uncles about it. "Don't worry," he said. "I know what to tell them."

"Will you let me know what happens?" I asked.

"Of course," he said. "In the meantime, go about your own business and let me handle it. Don't get into an argument with any of them. Don't blow your top, as that's just what they're looking for you to do. I know those girls. They were brought up like birds in a cage. They've never been anywhere. The only place they let them go alone is to a neighborhood show."

In a couple of days Vito called me and said that the Sicilians, meaning Mildred's uncles, had been down. He said he told them that they should keep their noses out of this matter. If they are fit to marry their wives, Joe is fit to marry Mildred. If Joe ain't fit, none of us are fit. Besides, he says, just so they know how he feels about it, he told them that he wants to be my best man. He told them that Charley knows about it and feels the same way. By Charley, of course, he meant Charley Lucky.

Then Vito said to me, "I'll do even better for you. I'll go up to the house and talk to the mother. Make an appointment for me." So I did, and Vito went up there and told her that he took full responsibility for me. That settled it, but the old lady still wasn't ready to make things any too easy. Mildred and me would have to wait six months before the engagement could be announced. She couldn't go out alone with me, and the only place I would be allowed to see her was at home and only on Sunday.

To tell the truth, I almost gave up. When I would go up there on Sundays, Mildred and me would never have a chance to be by ourselves. After eating, all we could do was just sit there and talk. The old lady or the brother was always around. If Mildred and me got too close sitting on the sofa, they always had some phony excuse to call her out of the room. I don't know how I lasted until the engagement party. That's when the wedding was set for September 18, 1932. Between the time of our engagement and the wedding, it was the same Sunday business, only now Mildred and her sister Rose were busy lining up bridesmaids, buying clothes, looking at apartments, and all that kind of stuff. Naturally, I was busy taking care of my slot machines.

The party after the wedding was at the Palm Gardens on 52d Street, right off Broadway. It was very large and cost close to $1,000 just for the hall. That was big money in them days; you must remember there were a lot of people in the street selling apples. I got two bands so that everybody could dance without stopping. For food we had thousands of sandwiches and about $500 worth of Italian cookies and pastries. There was plenty of wine and whiskey, even though it was Prohibition. I also got twenty-five barrels of beer as a present from one of the boys.

Now this was when the Maranzano killing and all the trouble before that was still in everybody's mind. There was a lot of friction right under the surface, so I had to weigh the invitations very carefully. But I must say it was a hell of a turnout of people either coming or sending money.

Vito Genovese couldn't make it to the wedding—I forget why—so he had Tony Bender represent him. But he made it to the party. Tom Gagliano and Tommy Brown were there. Charley Lucky sent an envelope. Willie Moore, Frank Costello, Joe Bonanno, and Joe Profaci also sent envelopes. The Raos came in person. Doc from the old group that used to be with Mr. Maranzano came, but Buster was dead by this time. Albert Anastasia and Carlo Gambino were there. So was Vincent Mangano, who was a boss then, but now has been missing for a long time. Joe Adonis sent an envelope. John the Bug came. So did Bobby

Doyle, Tommy Rye, Frank Livorsi, Joe Bruno, Willie Moore's brother Jerry, Johnny D, Petey Muggins, naturally The Gap, Mike Miranda, and all the boys with Tony Bender, which was my crew. It's impossible to remember everybody's name, but the hall was full.

After all the expenses of renting an apartment and buying the furniture, even an Oriental rug, we had about $3,800 left over from the envelopes of money we got. The only thing wrong was that the apartment Mildred picked was on Briggs Avenue in the Bronx near Mosholu Parkway—in other words, just a stone's throw from her mother.

Despite this romantic interlude, it was back to business as usual. Soon after his marriage to Mildred, Valachi was handed his first contract to kill since joining the Luciano Family. In the Cosa Nostra a soldier like Valachi was not paid for such an execution; it was simply part of his job. He did not know the victim and was only vaguely aware of the reasons why he had to die. His lieutenant, Tony Bender, relayed the order. The information was fragmentary. The marked man was known as Little Apples, and Valachi never did learn his real name. Bender said that he was about twenty-two years old and frequented a coffee shop on East 109th Street. Bender mentioned in passing that two brothers of Little Apples had tangled with Luciano and Genovese several years before and were slain as a result. Apparently there was some concern that he would now attempt to avenge their death. Valachi did not press Bender for details; he really could not have cared less.

When a soldier is given a contract, he is responsible for its success. He can, however, pick other members to help him carry it out. Valachi chose Petey Muggins and Johnny D, both of whom joined the Luciano Family at the same time he did. Then he began hanging out in the coffee shop and eventually struck up an acquaintance with Little Apples. During the next two or three days Valachi would drop in periodically for coffee and more conversation with him. Meanwhile, he scouted out various locations for the killing and finally settled on a tenement a block away on East 110th Street. For Valachi's purposes it was ideal. The ground floor was unoccupied, and there was no backyard fence to hinder a quick exit. His plan was first to station Muggins and Johnny D in the hallway and then to lure Little Apples to the tenement on the pretext that a crap game was going on in an upstairs apartment.

On the night of the hit, Valachi had arranged to meet Little Apples in the coffee shop. "Hey," he said, "let's take a walk. I hear there's a big game going on up the street."

"Great! I got nothing else to do."

According to Valachi, he positioned himself behind Little Apples as they were entering the tenement and suddenly wheeled away. "I heard the shots," he says, "and naturally kept walking down the street."

(New York police records reveal that about 9:20 P.M. on November 25, 1932, a male, white, identified as one Michael Reggione, alias Little Apples, was found in the hallway of 340 East 110th Street. Cause of death: three gunshot wounds in the head.)

Valachi went straight home. "After all," he recalls, "I was just married a couple of months, and I didn't want Mildred to think I was already starting to fool around."

6

Valachi's slot machine operation, so nicely protected by Frank Costello's stickers, fell apart not long after his marriage. Charges of municipal corruption had already forced Mayor Jimmy Walker to resign, and reform candidate Fiorello H. LaGuardia had vowed to run the machines out of the city. Even after LaGuardia became mayor of New York, Costello stubbornly fought to keep the racket going, and his political pull was such that he actually managed to obtain a court injunction restraining LaGuardia from interfering with the slot machines. But LaGuardia simply ignored the order and sent flying squads of police around town smashing them wherever they were found. A bit nonplussed at this "illegal" act, Costello eventually gave up and, at the invitation of Louisiana's Governor Huey Long, moved his slot machine base to New Orleans.

Valachi turned to pinball machines, but it was only a temporary measure. The best he could hope to realize from them was about $200 a week. He got the pinball machines from a jobber who suggested forming an association to control their distribution in upper Manhattan. Valachi was tempted, but still unsure of his position in the Cosa Nostra since the Maranzano murder, he was hesitant to strike out so boldly on his own. "After all I done to get rid of Joe the Boss and them other guys," he bitterly recalled during one of my interviews with him, "I had to walk with my head down because the old man went crazy trying to control everything."

Finally Valachi and Bobby Doyle asked Vito Genovese for permission to enter the already well-established numbers racket. The policy game, as it is sometimes called, is perhaps the simplest form of mass gambling ever devised. The name comes from the penny insurance that was being peddled in the late 1920s and early '30's; playing the numbers was just like taking out a cheap policy. All a bettor has to do is pick three numbers from 000 to 999 which make up the winning combination on a given day. The mathematical odds against this, of course, are 1,000 to 1, but the payoff is never more than 600 to 1 and often less.

How the winning combination is determined differs throughout the country—spinning a wheel, the daily dollar volume on the stock market, the total pari-mutuel handle of a previously designated racetrack, etc. When Valachi started, New York had its own version based on racetrack betting. The win, place, and show figures for the first three races were added, and the first digit to the left of the decimal point of the total became the first number of the winning combination; thus if the total was $189.40, "9" was the first number. The same procedure was followed in the fourth and fifth races to arrive at the second number, and in the sixth and seventh races for the final number.*

This staggered system of selecting the winning combination in New York allows a variation of the policy game called single action, in which a player can wager on individual numbers instead of all three. The payoff in this instance is 7 to 1. For a numbers banker like Valachi, single action betting had a special advantage. If, for example, the first two numbers of the winning combination were 58, he would quickly check his overall play as charted on a blackboard. "I only charted from a quarter up," he told me. "I didn't fuck around with dimes. Now I look for all my 58 leads. Maybe I'm lucky and don't have any. And maybe I got a 580 on the board, but it ain't being played heavy. So I take a chance and stay pat. But maybe there's a heavy play on 589 and I'm going to be crucified if it hits. I got to protect myself, so I bet a couple of hundred, whatever I have to, on 9 with another bank that handles single action. In other words, I'm edging off. I'm still hoping the 9 don't come up, but if it does, at least I'm breaking my fall."

For years the policy game had been a particular favorite of lower-income groups. Much to the annoyance of the Italian underworld, however, the celebrated Prohibition beer baron Dutch Schultz was the first to perceive the fortune that could be made by organizing it into a gigantic racket. Nowhere was it—or is it now—more popular than in Harlem. When Schultz moved in there, some thirty-odd policy banks were fiercely competing with

---

*Today this system is still prevalent in Harlem, where Valachi operated, although one may also play the so-called Brooklyn number based on the three digits immediately to the left of the decimal point in the pari-mutuel handle at a specific racetrack. In case anyone has money left over, there is, finally, a "night number," which is determined by the betting at trotting races.

one another. He put them together into one combine. His methods were simple. First, he terrorized individual bankers into paying him protection. Then, when he had them thoroughly cowed, he took over their business. "To keep the peace," as Valachi puts it, Schultz appointed Ciro Terranova, who was still the local Cosa Nostra power in Harlem, as a sort of junior partner. There were also two other major banks in the city formed by Willie Moretti and two brothers in the Gagliano Family, Stephano and Vito LaSalle, but the operation run by Schultz was by far the biggest.

Vito Genovese's blessing was necessary before Valachi could be admitted into the organized structure of the numbers racket. While some independent or, as Valachi says, "outlaw" banks remained, they not only were unable to participate in the fixed winning combinations the organization periodically arranged, but were often fleeced by these fixes. They could not, moreover, always take the advantage of edging off against potentially heavy hits or enjoy the general protection the organization afforded.

Even so, the going was not especially easy at first for Valachi. He had come into the Cosa Nostra too late to amass the kind of money from bootlegging that financed the move of so many other members into various industrial, labor, real estate, and gambling rackets when Prohibition ended. Although in the long run the odds are fantastically in favor of the numbers operator, there are inevitable losses that he has to absorb from time to time. Valachi learned this the hard way:

Bobby and me are only operating for maybe three weeks, trying to build things up little by little, hoping we won't get hit too bad. Well, this number pops that's got $12 or $14 on it, I think $14, and after we pay off our controllers and runners, we only had about $1,700 net that day and only a couple of thousand more in the bank. Now you can see how much we got banged for. We got hit for $8,400. That was a pretty good hit in those days. I don't know what I'm going to do. I call Willie Moretti and I tell him, "Tomorrow I'm bringing my wife over to your house. You support her."

Willie says, "What's the matter?"

"I'll tell you what's the matter," I say. "We got wiped out. We got nothing."

"All right," Willie says, "don't pay off. Stall your people."

So when all the runners come around that night, we tell them it ain't official yet, there's something wrong, and we got to wait for twenty-four hours.

Then, when the bets come in the next day, Willie calls and asks for a couple of our bad numbers—the ones with the biggest play on them. You see, he is going to try to lay them off for us so we don't get hit too hard and help pay off for the day before. But it ain't easy. He calls around to some other banks and tells them, "Listen, some of our friends got wiped out. How are these numbers for you? Can you handle them?"

Well, the other banks are getting too much play on these numbers, and we give Willie the next worst number. If it has to be, it has to be. Willie tells us he will handle it personally. He will accept $5 on the number to cover us, but not to make a habit of it. So we sweat it out and get through the day. But this ain't no good working without any money. We were lucky, but how lucky can we stay? The way it's going, me and Bobby ain't booking numbers, we're gambling ourselves.

Valachi's Family boss, Charley Lucky Luciano, bailed him out. Luciano's father had died, and Valachi went to the wake. While he was there, Luciano came over to him and said, "Hey, Joe Cago, don't look so sad."

It was a rare opportunity to speak directly to Luciano, and Valachi made the most of it. "Well," he replied, "I'm sad for your trouble, and to tell the truth, I'm a little sad for myself. I'm going broke in the numbers."

Luciano promptly instructed Valachi's old pal Frank Livorsi to provide him with some working capital. Livorsi gave Valachi $10,000. In return, Livorsi, Joseph Rao, and Joseph (Joe Stretch) Stracci became partners in the operation. Valachi and Doyle, however, were responsible for running it on a day-to-day basis, and to assist them, they took on a chief controller named Moe Block, who was an experienced numbers man. "You had to keep your eye on Moe," Valachi notes, "but he knew his way around and brought in some good controllers with him."

These controllers were vital to a successful numbers bank. For the most part Jewish, they were, in effect, branch office managers. As their title indicates, they also doubled as bookkeepers. This was one occupation that members of the Cosa Nostra rarely engaged in; accounting chores apparently were beneath their dignity, if not their ability.

Each controller had his own group of runners who actually collected the individual bets and returned any winnings. A controller got 35 percent of his daily gross if he was responsible for paying off the police in his area and 30 percent if the bank itself took care of this expense. The controller, in turn, gave his runners 20 or 25 percent of his share* "It was his business what he paid

---

*A runner also traditionally shares in a percentage of the winnings of one of his customers. If a player, for example, wins $600 on a $1 bet, the runner deducts $100 for himself.

them," Valachi says. "All we care is that their work comes in every day. If it don't, that's when we send for the controller."

From the time the numbers game received serious attention in the organized underworld immediately following Prohibition, it has blossomed into a racket currently thought to be taking in upwards of a quarter of a billion dollars annually in New York alone. It is unquestionably the chief source of police graft today. In a single precinct there may be fifteen or twenty locations—a tailorshop, a grocery store, a newsstand—where numbers bets are handled; to operate just one of them without police interference can require payoffs, to a patrolman walking his beat on up, amounting to $2,000 a month.

Nowadays bets usually start at a quarter, $1 bets are common, and $10 bets are accepted by the larger banks if they come from a regular customer. In the Depression years when Valachi began, nickel-and-dime bets were more the rule; even so, the numbers bank controlled by Dutch Schultz had an average daily play of $80,000.

Valachi's own operation, with the infusion of cash from Frank Livorsi, finally got off the ground. Another tangible benefit resulted from the new partnership when Livorsi informed him that they would share in a numbers fix. While the parimutuel figures at New York tracks could not be rigged, the racing season there was much shorter than it is now, and once it was over, the winning combination was derived from the betting at other tracks which the underworld had infiltrated. Among them, according to Valachi, was the Fair Grounds in New Orleans, Chicago's Hawthorne track, and Coney Island in Cincinnati. "The way I understood it," he says, "Dutch Schultz had this guy who could figure out how much money to put in the mutuel machines to make the right numbers come out. I can't remember his name. It was hard to say."*

Thus launched, the partnership built up a $60,000 bankroll with a daily betting volume of around $5,000. Out of this, as his share, Valachi began drawing a nifty, tax-free $1,250 a week. There were also the advantages of belonging to the organized structure of the racket in New York. Once, when he stopped to check on a controller, Valachi learned that the day's bets had not been picked up yet and volunteered to take them to the main office himself. Later, as he was getting out of his car with the bets and policy slips in hand, he was stopped by two detectives. Although he had been paying off the local precinct police, the detective turned out to be from the commissioner's confidential squad, and, as he notes, "You couldn't do business with them." He was arrested, but when his case came up for trial, his previous arrest record was not given to the judge, and he got off with a suspended sentence. "I don't know how it was done," Valachi told me, "and I didn't ask no questions."

(Valachi's file, according to the New York City Police Department, shows that he was arrested on January 13, 1936, on a policy charge and on September 12 received a suspended sentence in Special Sessions Court.)

Despite such prosperity and protection, when Valachi discussed his involvement in the numbers racket, he sounded more like a harried businessman than a rising racketeer. Except for the pure pleasure even he could not conceal while reminiscing about a fix—"I'll never forget that first number, 661; we won $7,000 and

---

*Valachi is referring to Otto (Abbadabba) Berman, a mathematical wizard Schultz used not only for fixes at racket-controlled racetracks, but also to manipulate mutuel odds so that the most lightly played numbers in his bank won.

had a swell Christmas"—he much preferred to dwell on all the
headaches of the operation. A constant cause for alarm were numbers that were regularly and heavily played. "We would get 222 a
lot and 725, don't ask me why," he recalls, "and always 000 from
the coloreds." Worse yet was the trauma he experienced when
some event of the day triggered a rush on a particular number:

We were using a New York track, so nothing could be fixed, when
there was this big payroll robbery in Brooklyn. I think it was at some
ice-cream company. Anyway they got away with $427,000. It was all
over the front pages about being the biggest job ever pulled.*

Now everybody has to play 427. We must have had around $100
on it. If it hits, that means we get banged for $60,000. Naturally we
can't pay off, as we ain't been in business that long. So I call this fellow who is with the LaSalle brothers to lay off the $100. But he,
speaking to me as a friend, says, "You are wasting the $100. If that
number hits the way it's being played, nobody can pay off. But I'll do
something for you. I'll give you $40 on it."

"Would you pay off?"

"For you, yes."

"Gee," I say, "thanks."

To make a long story short, it don't hit. But believe me, I took
notice of that number. The play on it kept getting weaker and weaker
as time passed. Maybe three or four years after this, 427 finally pops,
but not for much. I remember the bet. It was for fifty cents.

---

*On August 21, 1934, an armored truck on the Brooklyn waterfront near the
Rubel Ice Plant was held up by ten masked gunmen who made off with
$427,950. It was at the time the richest such haul in U.S. history.

Nothing, however, caused Valachi more distress than the attempts by both those in his employ and the world at large to, as he says, "clip" him or otherwise do him in:

You never knew what some people would be scheming next. Now my top controller was Moe Block, and one day he comes to me and says there are these guys out on Long Island who asked him if he was connected, meaning mobbed up. Moe says to me that he told them, "no."

"What did you say that for?" I say.

"I don't know," he says, "I just did."

Then Moe explains that these guys wish to come in with us. They want to turn in their work—which is what we call the envelopes with the bets in them—to us.

I ask Moe, "How far out on the island are they?"

"About three-quarters of an hour," he says.

I thought there was something fishy about the whole setup, but Moe says that just in case we think something is wrong, these guys want him to ride with them when they pick up the bets.

"Well," I say, "I still don't like what I've heard so far, but we'll see. If I find out I'm right, we won't pay off."

You got to remember that the first number in a winning combination comes from the first three races at whatever track we're using. What I don't like is that by the time Moe is back in the city with these guys, the first three races are going to be over. But we can always use more business, and I tell Moe we'll try it.

Well, Jesus, it happens. The first day they hit us with a $3 bet, and that's a $1,800 payoff. Right away I tell Moe to chart all this new work by itself. Now, if the first number in a winning combination leads off with a 4, there are only 100 possible ways to arrange the next two numbers—from 400 to 499. When I look at the board after Moe is finished charting, I see that in the work from these guys on Long Island, all 100 possible combinations leading off with 4 are up there. Of course, there are some other bets mixed in, so we don't get suspicious, but not enough when you saw them all lined up at once.

Now I wait for them to come and collect the money, but they call up instead. Before I can say anything, they are already yelling, "We didn't know you was connected! We thought you was independent. We just found out." Then they say to please not go after them. I say, "Why not?" and they say that after all, they asked Moe if he was connected, and if they had known the truth, they never would have done it.

Naturally I'm interested in learning the way they operate—who knows who else will try it?—so I say, "Okay, if you tell me how you clipped us."

"Well," the guy says, "by the time we are coming over the 59th Street Bridge, the first three races are in, and we know the first number is set. We find out what it is after we come off the bridge. There's a barbershop on the right, and one of our guys steps out of it and holds up four fingers as we go by. Now we've got ten envelopes, each leading off with a different number from 0 to 9. So when we get the high sign from our guy in the barbershop, we just give Moe the envelope with all the combinations that start with 4."

Can you imagine the nerve of them guys?

Moe Block also figured in an infinitely more ominous development that might have abruptly ended Valachi's underworld career had he not been backed by the Cosa Nostra. Among the controllers Block had brought with him in the numbers operations was a "guy named Shapiro." Shapiro, according to Valachi, was a valuable addition, since he regularly turned in between $300 and $400 a week. "That," he points out, "was big money then. You got to remember times were bad—what they called the Depression was on—and most of the bets was nickel-and-dime stuff. Today it's $1 and more. So you see Moe was impressed with Shapiro. In other words, this Shapiro had him bulldozed."

Shapiro apparently had ambitions to be more than a mere controller, and joined "two Jew boys," Harold Green and Harold Stein, in an attempt to muscle in on Valachi's numbers racket. Valachi had already heard of Green and Stein. "They are from around the Bronx somewhere," he says, "and they think they're pretty tough. Their gang has been sticking up numbers runners, bookies, everything you can think of." Valachi first learned of their interest in him when an obviously nervous Moe Block passed the word that Green and Stein had been asking "around" about him. "They mean business," Block added. "They are set to take over the city."

As was the case with anyone who worked for him, Valachi had said very little about himself to Block and nothing, of course, about the Cosa Nostra. "Moe only knows what I tell him," he says. "I just said to him that I can handle any trouble that comes along, and he should never worry. So I tell him to stop talking about how tough these guys are, or I'll start thinking things about him." Then he asked Block what Green and Stein knew about him.

"They just know you're in the numbers with Bobby Doyle."

"Who tipped them off?"

"Shapiro," Block said.

Valachi recalls that he was not especially surprised at the news. "This Shapiro," he says, "was acting pretty big. A couple of days before he come up and told me how much he likes me. I just said, 'Yeah?' but I was thinking, where does *he* get off telling me he likes me."

Valachi immediately alerted the three silent partners he and Doyle had—Frank Livorsi, Joe Stretch, and Joseph Rao:

Well, the next thing we know, two guys from this outfit walk into our main office in Harlem. It happens that Joe Rao is the only one of us there at the time, and these guys want to know how Bobby Doyle and me fit in with him. "We are all partners in the business," he says. So they tell him they want to give us protection for $500 a week.

"From what?" Joe says. They say there could be all kinds of trouble, you never know from where, and they would handle it for us. Joe goes along with them and asks when they want the first payment. They say they will be back for it on Saturday. We all got a good laugh when we heard this. If they had come, it would have been a slaughter. But they don't show up.

So now Joe Rao decides to put the snatch on Shapiro. When they got him, they tell me to get over there and talk to him. They had him in the back of a truck in this garage, and he was beat up good. His skull was busted, and there was blood all over the place, and there was a rope around his neck. I think Jimmy Blue Eyes, right name Vincent Alo, was holding the rope.

As soon as Shapiro sees me, he yells, "Save me! Save me!"

Was he kidding—save him? What could I do? They were asking him if Moe Block was in on the deal. I tried to find out myself, but the way Shapiro was yelling anything and everything when he got a jerk on that rope, I couldn't make out if Moe was in it or not. Well, that was the end of this Shapiro. He was put in a drum full of cement—the kind they use for oil—and dumped in the East River. I guess he's still there.

I kept Moe on because a good controller was hard to find. Even if he was in with them, he learned a good lesson, and I figured he wouldn't be pulling nothing again. Right?

Valachi was doubly thankful for his Cosa Nostra membership when he found out that the man behind Green and Stein was

Dutch Schultz. For all his power, Schultz was necessarily ignorant of internal Cosa Nostra affairs. "When The Dutchman took over the numbers in Harlem," Valachi says, "he had cut in Ciro Terranova and the boys on 116th Street. Now all he knows about me and Bobby Doyle is that we used to be with Maranzano against Ciro and Charley Lucky and Vito. So when he hears about us being in the numbers right under his nose, he figures we are just outlaws." Since Joseph Rao was widely known to be an associate of Terranova's, Schultz was understandably put out when Green and Stein reported back to him that Rao was also a partner in Valachi's numbers bank. As it turned out, it was simply another gambit in Luciano's strategy to make the Cosa Nostra supreme in the U.S. underworld:

The way I understood it, The Dutchman goes to Ciro and wants to know what the hell's going on. Now Ciro is up a tree as he don't know about Joe Rao being with us either. Then he finds out from Joe Rao that we have the okay personally from Charley Lucky and Vito to book numbers. So Ciro calls Charley Lucky, and Charley says to tell The Dutchman that we are being allowed to build up our business in order that they, meaning Ciro and The Dutchman, can take it over. Naturally Ciro has to do what Charley orders. When I hear all this, it don't take much for me to see that Charley Lucky has something in mind for The Dutchman and that Ciro Terranova, who is supposed to be such a big shot, ain't in on it.

I am right. Right after this Vito Genovese tells me and Bobby Doyle that The Dutchman has got to go. Vito says the Jews have agreed. He says not to go out looking for him, but to shoot if we happen to bump into him somewhere. In other words, the thing is not to tip him off.

Dutch Schultz was a natural target for Charley Lucky Luciano. As a practical necessity, Schultz would develop a working alliance in the Harlem numbers racket with someone like Terranova; to do otherwise meant taking on the entire Italian underworld. But he remained stubbornly apart from the overall coalition Luciano was forming with such groups outside the Cosa Nostra as the so-called Bug-Meyer mob, run by Meyer Lansky and Benjamin (Bugsy) Siegel, and the gang headed by Lepke Buchalter and Gurrah Shapiro. Personally Schultz had become just as much an anathema to Luciano; he was the complete gangster caricature—flashy, boisterous, insatiably self-centered. Even more galling was the fact that Schultz, for all his crude ways, possessed a considerable talent for spotting and organizing new business opportunities. At the peak of his career as a Prohibition beer baron, he not only was one of the first to see the potential of the numbers game, but had also organized the immensely profitable restaurant racket. Schultz first strong-armed his way into control of the waiters union. Next he formed the Metropolitan Restaurant and Cafeteria Owners Association. If a restaurateur wished to avoid a strike or perhaps a stink bomb during lunchtime, he joined the association—at a price.

The idea of dividing up the Schultz empire had always appealed to Luciano. But the project had to be handled with some diplomacy. Schultz was nobody to treat lightly. Born in the Bronx as Arthur Flegenheimer, he quickly earned a reputation as a bully-boy. As a result, he was called Dutch Schultz after a local brawler who was active around the turn of the century. The alias stuck, much to his subsequent dismay when even he recognized that he was getting too notorious. Once he was heard to complain that it was because his adopted name made life easy for newspaper headline writers. "If I had stuck with Arthur Flegenheimer," he observed, "nobody ever would have heard of me."

By the late 1920s Schultz had absolute control of beer distribution in the Bronx. The methods he used to get to the top were incredibly brutal. The last Prohibition beer wholesalers standing in his way were two Irish gangsters, John and Joseph Rock. John Rock, after sizing Schultz up, decided to retire. His brother, however, elected to fight it out. He was promptly kidnapped by Schultz and his men. They beat him up so badly that he became a permanent cripple. It was just a start. Before they dropped him on a Bronx street, they smeared pus on a strip of gauze and taped it tightly to Rock's eyes. Eventually he went blind. Such tactics smoothed the way for Schultz into his other ventures. Any numbers banker, restaurant owner, or union leader who felt like protesting always had Joseph Rock as an example to reflect on.

Luciano's initial chance to move came when Schultz was indicted on the same kind of income tax charges that sent Al Capone to prison. Schultz went into hiding for some eighteen months while his lawyers vainly attempted to resolve the case. During the time Schultz was a fugitive, Luciano dickered with his chief lieutenant, Abe (Bo) Weinberg, about adopting a more cooperative attitude than his boss. Weinberg was amenable. Even when Schultz finally surrendered, it made little difference; the evidence against him seemed airtight. But Schultz, with some lavish spending in the right places, was acquitted.*

---

*At his trial in the tiny upstate town of Malone on the ground that he could not get a fair hearing in New York City, his biggest coup in influencing the jury was supplying children in the local hospital with flowers and candy.

The first to suffer the consequences was Bo Weinberg. Valachi heard about it from Bobby Doyle. "Remember that Weinberg you thought was such a nice guy?"

"Yeah," Valachi countered, "so what?"

"Well, he's dead and buried. The Dutchman done it."

"Hey, that's why I ain't seen him around lately," Valachi said. "What happened?"

"The word is that The Dutchman found out he was playing around with the Sicilian."

Weinberg's body was never found. He simply vanished. According to Valachi, he was carried out of a midtown Manhattan hotel in a trunk after a heated confrontation with Schultz. "When I heard all this," Valachi told me, "I figured The Dutchman wasn't long for this world himself."

For a time an uneasy truce prevailed between Luciano and Schultz. Then the government, refusing to let Schultz settle up his back taxes, began preparing another indictment against him. A New York City grand jury also began looking into his connection with the numbers racket, and an up-and-coming Justice Department attorney named Thomas E. Dewey was brought in as a special prosecutor. Schultz was furious. One of his own lawyers had warned him that if something wasn't done, Dewey would "indict us all."

One night Valachi got an intimate glimpse of Schultz's state of mind. He happened to go into a restaurant called Freddie's Italian Garden, on 46th Street west of Times Square, just as Schultz was leaving a dinner conference with Luciano. Luciano remained behind, and Valachi was later invited to join his table. When someone in the group mentioned Schultz, Luciano smiled and remarked, "All The Dutchman can talk about is Tom Dewey this and Tom Dewey that."

Schultz's solution to his problem—assassinating Dewey—gave Luciano the chance he had been waiting for. He found almost universal agreement among underworld chieftains that if Schultz were allowed to murder Dewey, it could trigger exactly the kind of all-out drive against organized crime that they were anxious to avoid. Thus Schultz himself had to be liquidated. It was about this time that Valachi heard from Vito Genovese that Schultz was to be killed on sight. But in the end, three of Lepke Buchalter's gunmen were specifically given the contract. As Valachi noted to me, "Charley Lucky figured it was best all around that The Dutchman's own kind took care of him."

On October 23, 1935, Schultz was shot down in a bar and grill in Newark, New Jersey. He did not die at once. A Jewish convert to Catholicism, he received the last rites in a local hospital from a priest while a police stenographer recorded his final fever-ridden remarks; but they read like lines in the mad scene from *King Lear*. For someone who fancied himself a tough character his last coherent sentence was a pathetic moan: "Let them leave me alone."

Buchalter inherited Schultz's restaurant racket, and Luciano and Genovese took over the huge Harlem numbers bank they had eyed for so long. "Charley Lucky opened the numbers up for everyone in the Family," Valachi says, "except of course you couldn't go into another member's territory or take his controllers and runners." To oversee it, Valachi's ancient enemy, Ciro Terranova, was replaced by another Luciano lieutenant, Trigger Mike Coppola. The erstwhile Artichoke King did not protest; getting on in years and in poor health, he was glad to go into retirement. As Valachi says with a measure of regret in his voice, "Ciro was able to die in bed."

To all intents Luciano's strategy was flawless, and a sense of well-being filtered throughout the Cosa Nostra. Valachi's numbers operation prospered to the extent that even he did not mind the additional "ice," or police payoffs, which expansion entailed. Out of his profits he financed a "classy horse room" in White Plains, New York, where suburban matrons could while away their afternoons betting on races across the country. He also began to dabble in a little loan-sharking on the side. "If you ask me how I was doing," Valachi recalls, "I would have to say I was doing okay."

But Dutch Schultz got in a last snicker from beyond the grave. The public had long been indifferent to, if not amused by, racketeering; indeed, a Luciano or even a Schultz was as much a prize for a hostess of the day as the latest best-selling author or matinee idol. Suddenly, as occurs periodically, a tremor of righteous indignation swept the country. It could not have happened at a worse time for Luciano. In New York Special Prosecutor Dewey, hard-nosed, politically ambitious, a Republican not at all shy about poking into a corrupt Democratic machine, was out to make a name for himself as a rackets buster. Deprived of Schultz as a target, Dewey zeroed in on Luciano, ironically the man who probably saved his life, and he would use a surefire subject—vice—to rivet everybody's attention on the case.

Valachi picked up a hint of the trouble to come during a rare visit to a Manhattan bordello. "I didn't go in for this," he says, "but one night some of the guys wouldn't let me say no." While he was there, he overheard one of the whores whispering fearfully that they were Charley Lucky's boys and to be careful. Valachi did not think Luciano would especially appreciate his name being bandied around like that, and he intended to report the conversation. Before he could, however, Luciano was arrested in Hot Springs, Arkansas, at Dewey's request and returned to New York to face trial on multiple counts of compulsory prostitution. "Well," Valachi told me, "I was stunned. Charley Lucky wasn't no pimp. He was a boss."

In a way Valachi was right. Luciano really did not need prostitution as a source of income, but he got into it precisely because of the responsibilities of his position. When the Prohibition bonanza was over, a certain amount of unemployment resulted in Cosa Nostra ranks, and the search for new rackets was at a premium. At the time vice in New York was based more or less on the free enterprise system; upwards of 200 independent brothels were operating in the city and doing a handsome business. For Luciano it was simply a question of organizing them into a cartel himself or letting someone else beat him to it, as Dutch Schultz had done in the numbers game.

Luciano, as history has recorded, was sentenced on July 17, 1936, to thirty to fifty years in prison. Even worse for a man of Charley Lucky's fastidious tastes, he ended up in Dannemora, a bleak, maximum-security penitentiary near the Canadian border not inaccurately known among convicts as Siberia. There he remained until 1942, when his controversial role during World War II began.

The Navy had become alarmed about possible sabotage and intelligence-gathering activity along the New York waterfront by enemy agents. Somehow the idea took hold of enlisting the organized underworld on the docks to prevent this, and what better intermediary to arrange it than the top man himself? Luciano's first tangible benefit for his cooperation was being transferred from Dannemora to a more accessible and comfortable prison just outside the state capital at Albany. Just what he actually did has never been pinned down. Even more debatable was what followed. In 1945 Luciano's lawyers pushed for his parole because of his contributions to the war effort. It was eventually granted. Closely tied in with this decision was the fact that he had never bothered to take out U.S. citizenship, and once paroled, Luciano was promptly deported to Italy on the fallacious theory that he would cease to be an American problem.

Dewey's successful prosecution of Luciano produced shock waves in the Cosa Nostra that a lesser organization might not have survived. Not since Capone had such a potent figure in its ranks been toppled. Charley Lucky's fall, moreover, was far more dramatic. Capone, after all, was a crude loudmouth, while Luciano with his urbane veneer epitomized the "new racketeer," anxious to be portrayed as just another business tycoon pushing buttons from behind his mahogany desk.

For Valachi it was a personal loss of considerable magnitude. "Gee," he says, "I thought why did it have to be Charley Lucky? Why couldn't it be somebody else? Everything was going swell with me. He helped me a lot in the numbers, and I was getting closer to him. Who knew what was coming up next? Believe me, I felt terrible. Every time somebody took an interest in me, I was deprived of him."

Luciano's departure from the scene would also affect Valachi in a way he could not possibly have imagined. It opened an avenue to power for Luciano's second-in-command, Vito Genovese, the man who some twenty years later would sentence Valachi to an underworld execution. At the time, however, nothing seemed less likely. While Luciano allegedly was doing his patriotic best for the

United States, Genovese was in, of all places, Italy, where it later turned out he became such a darling of the Fascist regime that Mussolini himself had decorated him with the highest award he could bestow on a civilian.

Genovese is a man of Byzantine bent. "If you went to Vito," Valachi says, "and told him about some guy who was doing wrong, he would have this guy killed, and then he would have you killed for telling on the guy." Born on the outskirts of Naples in 1897, he was educated there through the equivalent of the fifth grade. Genovese arrived in this country when he was sixteen, and his earliest known haunts were in New York's Greenwich Village, an area favored at the time by Italian immigrants, as well as artists and writers. He was married at some point prior to 1924, but his wife died in 1929.

Later he met the love of his life, Anna Petillo, who was unfortunately married. Then, as luck would have it, her husband was murdered, and within two weeks Genovese married her in a civil ceremony witnessed by Anthony (Tony Bender) Strollo and his wife. According to Valachi, Peter (Petey Muggins) Mione and Michael Barrese killed him on Genovese's orders:

Now this is back in 1932, and I'm close to Petey Muggins. Remember he is one of the guys who went with me to Charley Lucky and Vito after Mr. Maranzano got his. One day we're talking about this Mike Barrese from the Village who is hanging around in Harlem where we are. So Petey says it's on account of him and Barrese strangling a guy on a roof downtown on Thompson Street.

Petey says some people saw them doing it. It's okay, they have been straightened out, but this Barrese can't stop being worried, and he's staying up in Harlem, where nobody knows him. In other words, he's afraid to go downtown and be seen. Petey says that ain't so good. He should go back and let the bulls question him if they want to, and the whole thing will blow over.

"I don't know nothing about this," I say. "Who was the guy that got killed?"

Petey looks around nervouslike, and he says to me, "Joe, don't tell a soul. It's only on account of you being you that I would tell you. The guy was Anna Genovese's husband."

"Eh," I say, "so that's how things are."

Well, this Barrese disappeared. I never saw him again. I don't know what happened to him, but it ain't hard to figure.

(According to New York City police records, one Gerard Vernotico, age twenty-nine, of 191 Prince Street, was found dead at 2:15 P.M., March 16, 1932, on the roof of a building at 124 Thompson Street. Vernotico's widow, the former Anna Petillo, married Vito Genovese twelve days later in the Municipal Building, Manhattan. Vernotico's arms and legs were bound, and a tightly drawn sash cord was around his neck. Also found dead with Vernotico was one Antonio Lonzo, age thirty-three, of 305 East 28th Street. It is believed that Lonzo was killed because he was a witness to the slaying of Vernotico. Case open.)

At least Genovese's love for Anna seems to have been undying. Even when she testified in 1950 during divorce proceedings against him about his racket connections and the size and sources of his income—an unheard-of act for anyone who valued his health—she remained unscathed. "Nobody could understand why Vito didn't do anything about her," Valachi recalls. "The word was all around, why don't he hit her? But he must have really cared for her. She had something on him. I remember when we—Vito and me—were in Atlanta together later on, he would sometimes talk about her, and I would see the tears rolling down his cheeks. I couldn't believe it."

If so, she is Genovese's only discernible soft spot in a brutal career. During the 1920s he was periodically arrested on various charges, including homicide and felonious assault. He had, however, only two convictions, both for carrying a concealed weapon—a pistol. The first, when he was twenty, got him sixty days in the workhouse; the second time he escaped with a $250 fine.

After that, as he loomed increasingly large in the Cosa Nostra, he generally managed to stay out of the way of the law with one apparently minor exception at that time. In 1934 he and a Luciano lieutenant, Michele (Mike) Miranda, bilked a gullible merchant out of $160,000 in two stages—initially in a crooked card game and then in that hoary flimflam, a machine that supposedly made money. To their intense annoyance, one Ferdinand (The Shadow) Boccia began to pester them for the $35,000 he had been promised as his share for enticing the victim into their clutches.

Boccia should have left well enough alone. He had gotten involved in the swindle in the first place trying to get back in Vito's good graces after he and a sidekick named William Gallo had held up a liquor store which belonged to a friend of Genovese. To catch Boccia off guard, it was decided to use Gallo and another small-time hoodlum, Ernest (The Hawk) Rupolo, to dispatch him. Then Genovese and Miranda got a bit too devious for their own good. They gave Rupolo $175 to kill Gallo, once Boccia had been taken care of.

Apparently Genovese and Miranda had some second thoughts about the whole thing and ended up assigning the contract to Cosa

Nostra professionals. A ludicrous sequence of events followed which doubles Valachi over with laughter every time he thinks about it. When Rupolo heard that Boccia had been murdered, he proceeded with phase two of the original plot. He and Gallo attended a movie in Brooklyn one night, and as they walked down the street afterward, he took out a pistol, put it against Gallo's head, and pulled the trigger: The pistol misfired. Rupolo quickly tried again. Still nothing. When Gallo demanded to know what was going on, Rupolo lamely passed it off as a joke and said that the pistol was not loaded. The two continued on to a friend's house, where Rupolo examined the pistol, discovered that the firing pin was rusty and oiled it. Upon leaving the house, they walked together for several blocks, and then Rupolo took another crack at Gallo. This time the pistol went off, but all Rupolo managed to do was wound him. Rupolo, identified by Gallo as his would-be assassin, was sentenced to nine to twenty years in prison.

There for a moment the matter lay. Some rumors circulated about those involved in the murder of Boccia, and Genovese was brought in for questioning. But little came of it. Then a more threatening situation developed. After Dewey had convicted Luciano, he named Genovese the new King of the Rackets and began probing into his affairs. All in all it appeared to Genovese to be an excellent time to drop out of sight for a while. He had visited Italy for three months in 1933, presumably liked what he found, and skipped back there in 1937.

According to Valachi, Genovese told him in the federal penitentiary in Atlanta that he had taken $750,000 with him when he fled to Italy. The figure is not far-fetched. Genovese's wife testified in her divorce suit that he had large amounts of cash in various

European safe-deposit boxes, including $500,000 in one in Switzerland, and after he was located in Italy near the close of World War II, it was discovered that he had donated $250,000 toward the construction of a Fascist party headquarters.

"Well," Valachi told me, "with Charley Lucky gone and then Vito, I must say it was a shock to all of us."

7

With both Luciano and Genovese at least temporarily out of the picture, Frank Costello became acting boss of the Luciano Family. This would have immediate and serious repercussions for Valachi. Not that he had any personal difficulty with Costello. "Frank," he notes, "was a peaceful guy, a diplomat." The trouble was that he did not know Costello well, and this left Valachi vulnerable to the whims of his lieutenant, Tony Bender, whom he had intensely disliked from their first meeting.

Costello, moreover, was far less concerned about Family matters than he was in caring for his own booming enterprise—his slot machine racket, which had become national in scale; his huge bookmaking operation; his gambling casino interests, including the famous Beverly Club outside New Orleans; and his partnership, necessarily silent since he was a former bootlegger, in a wholesale liquor firm then realizing $35,000 a month as the exclusive distributor in the United States for King's Ransom Scotch.

To inquiring reporters Costello liked to emphasize his legitimate interests in real estate, oil, and similar ventures, and he is in fact probably one of the few Cosa Nostra chieftains who might have achieved great success as a businessman. He also is not without a sense of humor. Once pressed about his role in gambling, he observed, "Some people are common gamblers. I am an uncommon gambler."*

When Costello wanted to, however, he did not hesitate to show his hand. His habit was—and still is—to enjoy the steam baths at a Manhattan hotel in the late afternoon whenever he could. The night manager approached him on one such occasion and explained that other clients were expressing some dismay at his presence.

"You mean you don't want me to come here anymore?" Costello said.

"If it were up to me," the night manager said, "you could come all you want. But we have been getting these complaints. You know how some people are."

The next morning none of the hotel's employees—chambermaids, waiters, elevator boys, maintenance men, kitchen help, and so on—reported for work. Eventually the frantic general manager discovered what had happened and immediately telephoned Costello.

---

*While writing a magazine article a few years ago, I had dinner with him. During the meal a lackey brought him an evening paper that featured a front-page story on Costello. I asked him if he ever read anything new about himself, or was it always the same old stuff? "Well," he replied in the gravelly voice that became so famous during the Kefauver crime hearings, "you won't believe it, but it's the God's honest truth. I never read nothing about myself. You know why? It only upsets my stomach."

"What are you telling me for?" Costello replied. "I don't have anything to do with the unions."

"I know that, Mr. Costello. I was really calling you to say that an unfortunate error was made last night."

"You mean I can use the baths?"

"Anytime you wish, sir."

Within hours the missing employees were back on the job.

Costello preferred to use political connections, instead of muscle, to advance his fortunes. But to hear him tell it, he had no more influence than any other man who lived in one neighborhood for many years. This is the same Costello of whom a Tammany Hall leader admitted, "If Costello wanted me, he would send for me." When Governor Huey Long was assassinated shortly after he invited Costello to bring his slot machines to New Orleans, it bothered Costello not at all; he simply cut Long's political heirs in for a share of the loot. Sometimes Costello's political power got a bit embarrassing for him. One memorable instance occurred when Costello backed the appointment of Thomas Aurelio to the New York State Supreme Court. At the time there was an authorized wiretap on Costello's telephone. Among other items it picked up was a call from Aurelio fervently thanking "Don Francesco" and pledging future fealty. Of his uncanny ability to persuade political figures to see things his way, Costello would explain, "I know them, know them well, and maybe they got a little confidence in me." Former Mayor William O'Dwyer offered a more practical solution. Asked what he considered to be the basis of Costello's appeal to politicians, O'Dwyer said, "It doesn't matter whether it is a banker, a businessman, or a gangster, his pocketbook is always attractive."

Such high-level wheeling and dealing did not leave Costello

much time to administer Cosa Nostra affairs, even if he wanted to. Thus by default the various lieutenants in the Luciano Family — among them Willy Moore, Anthony (Little Augie Pisano) Carfano, Joe Adonis, Trigger Mike Coppola, Dominick (Dom the Sailor) DeQuatro, and Anthony (Tony Bender) Strollo — enjoyed vastly increased power. To complicate matters further for Valachi, besides Bender hanging directly over his head, his relations with his partner in the numbers game, Bobby Doyle, had begun to deteriorate:

Little did I know that I was drifting off with the worst troublemaker in the world. I'm talking about that dog, Bobby Doyle. No one on earth can match him for his treachery except Tony Bender, and I blame Vito Genovese for making Tony think he is such a big guy. They are all dogs to me. I'm sorry I ever got mixed up with them.

When Bobby got me to go with Vito Genovese after Mr. Maranzano got killed, Vito was the underboss in Charley Lucky's Family. Vito put me and Bobby in the crew of Tony Bender. This made Tony Bender my lieutenant. His word is law now. Who is there for me to go to? There is Frank Costello, but I don't ever see him. All he is interested in is making money. I got nothing personal against Frank, but I must speak the truth. I curse the day I didn't go with Tommy Brown, right name Lucchese. He was a good friend of Mildred's father, and he would have looked out for me.

Right away I could see that Tony Bender was going to be trouble for me. It didn't take me long to see I was right about him. One day Bobby Doyle comes uptown from where he has been in the Village with Tony and tells me the Wacky brothers have cracked Eddie Starr's head open over a girl. Now Eddie Starr's real name was Eddie Capobianco. He picked up the name when he was going out with a dance-hall girl called

Mary Starr. Anyway Bobby said, "Do you know the Wacky brothers?"

"Yeah," I say, "they're in the numbers working for Vince Rao."

"Okay," Bobby says when I tell him I know the Wacky brothers, "Tony wants you to go after them."

Now the Wacky brothers and Vince Rao are in Tommy Brown's family, and I say, "Bobby, you know how I stand with Tommy. How come I'm picked? So hard feelings will come between us?"

"Relax," he says, "you're not going to kill them."

"What am I supposed to do?"

"Just work them over."

What could I do? I got to hope that Tony Bender knows what he is doing. So I say, "Okay, where are they?"

"I don't know," Bobby says, "you have to find them."

Well, I figure if I can't find them, I can't find them. But the next morning I get a call from Fat Tony Salerno. Fat Tony asks me if I am looking for the Wacky brothers. I say I am, and he says, "Which one?"

I say, "Who cares? I was just told the Wacky brothers."

Then Fat Tony says that he was going to meet one of them on business at 97th Street and Third Avenue at two o'clock in the afternoon. He says that he would be late, so I'll have a chance to get him. Now they have me, you see. I can't get out of it by saying I can't find the Wacky brothers. They must have put Fat Tony up to it.

They have already told me who I'm to use to help me. One was Johnny D. He is an enforcer for Tony Bender when anybody held out on him in the numbers. Johnny was as stupid as they come, even if I was the one who got him mobbed up. He had no teeth, and one day Vito Genovese said to him, "The next time I see you, I want to see you with some teeth." Afterwards Johnny tells me, "See, that shows he's interested in me." The other one was Tommy Rye. His real name is Eboli, and the bulls have him down as also Tommy Ryan, but I always knew him as Tommy Rye. He's a big shot now; in those days he was just a punk. Anyway, they each got a baseball bat. I don't have to have one, as I am in charge. We drive around for a while, and then we park the car on East 97th Street.

He shows up, just like Fat Tony said, but he is with another guy. We all jump out of the car. He sees us and freezes. Well, Tommy and Johnny D go to work on him while I hold back the other guy. I tell this guy, "Mind your business. This son of a bitch hit one of the boys over the head last night."

Usually you just go for a guy's legs, but Tommy and Johnny were hitting him over the head and on every part of his body before I could stop them. If you want to know what the people in the street were doing, they were just going about as if nothing was happening.

I don't think he yelled once. He just tried to cover up as best he could. Then he fell down, and I called them off. He had enough. This was just a beating, not a killing. If he dies, it isn't because it was meant to be. I heard he was in the hospital for about six months.

Now I am in trouble with Tommy Brown because the Wacky brothers are under his protection like I explained, and Tony Bender ain't cleared nothing. Instead of me going after one of them the way Tony made me do, they should have been called "to the table," which is like a hearing before a real trial. If there was going to be any trouble, Tony Bender was making sure it would fall on me. I find this out when Bobby Doyle calls me and says I got to take the rap. "Why?" I say, and Bobby says that Tony is in up to his neck about it. I ain't sad about that, but it don't do me no good either.

Well, I am lucky that Tommy Brown was the best friend of my wife's father, God rest his soul, because now I'm at the table for what happened. But it's almost Christmas, and everything is put off until after the holidays. Tommy used to have us over at his house for a party every Christmas, and

I am amazed when Mildred tells me that Tommy's wife, Kitty, called up and invited us again. Naturally I report this to Tony. I still remember his number. It was Cliffside 6-7835 or 3578, one or the other.* Tony says it's a good sign and to go, as if I need him to tell me that.

When we got there, I have a couple of drinks, and then Tommy calls me upstairs and asks me who ordered the beating. What could I say? Tony Bender was my own lieutenant, and if I talk, I would be in a lot more trouble. Maybe it would even start another war. So I say, "Tommy, I did it on my own. Where will I land if I tell you otherwise?"

He says, "Listen, I can break Tony Bender."

"Tommy," I say, "I'll tell you something. I understand what you're trying to say; but let's put it that I did it on my own."

Then he just made a motion with his hands which meant okay. After the holidays they had the table. It was at the restaurant Charley Jones had on 14th Street. Tommy was there, and so was Vince Rao. For me there was Tony Bender and the Family counsel, Sandino. Sandino was a greaseball, but he had a wise head. Whenever there was a table, he sat with Tony to make sure Tony don't make no mistakes.

I'm sitting off to one side while they talk over my case. Bobby Doyle was keeping me company, and he whispers, "Don't worry if things don't go our way." For once I think he was telling the truth, so I am twice as glad I didn't get Tommy Brown steamed up. Well, Tommy don't press the matter, but I must say his boys were always cold to me after that. Then everyone gets up and says good-bye, and Tony Bender comes over and says, "Everything is all right"—meaning the thing won't go no further.

It wasn't the last time Tony would cause me trouble.

---

*According to the Bureau of Narcotics, Bender's unlisted home telephone number in New Jersey was Cliffside 6-3578.

For a time after this Valachi continued a shaky partnership with Bobby Doyle in the policy game. He began to branch out, however, into other areas. The first of these was "shylocking," or the loan shark racket—the lending of money at high interest rates to borrowers who are unable or unwilling to obtain it through legitimate channels:

I'm fed up with Bobby, but I ain't going to give up the numbers just because of him. I figure I'll just see what happens and start looking around for other things. With Vito gone and Tony around, I can see I'm going to have to depend on myself. I'm not saying Vito was any prize, but at least he had some sense then. One thing is sure. I am going to stay away from Tony Bender. As he hangs out in Greenwich Village, naturally I stick to Harlem. The only time I go downtown is when I'm called.

Now I take some of my profits from the numbers and go into shylocking. The loans went for 20 percent interest, which is what we call vigorish. Take an example. You loan out $1,000 and the guy is supposed to pay back $100 a week for twelve weeks. The $200 you make is the vigorish. Figure it another way. For every $5 you lend out a week, you get back $6.

How did I start? Well, you make one or two loans, and everybody wants a loan. Jesus, if you gave to everybody who wanted money, you'd have to be the Bank of Rome. The word gets around the neighborhood. I am known in the Bronx, where I live, and I am known in Harlem because of the numbers, so that's where I dealt. Naturally you give the ones you feel are more solid.

Sometimes I'm stuck, I don't have enough cash, and I go to a shylock myself. He would charge me 10 percent, while I am charging my people 20 percent, because he knows it is a solid loan. He knows he is going to get his money back from me. He ain't taking a risk.

There has been a lot of stuff in the papers and whatnot about shy-locking, meaning all the rough stuff. All I can tell you is how I worked it. Some guys did it their way. I did it mine. I tried to run it as a business. I'm not looking to beat up somebody. I want to make money, and the idea is to keep it circulating. What good is it laying on the ground? If you don't believe me, ask around. The truth is that I never got picked up for shy-locking, and I was in it for years, not always steady, but I made a living out of it. I wasn't too big. After all, as you will see, it was a thing I had going on the side. In other words, it wasn't my main business.

I didn't lose a penny. I always collected on every loan. The reason is I was careful who I dealt with. I didn't do much business with busi-nessmen, you know, legitimate guys. A businessman after a while starts to think about all the interest he's paying. After one loan he comes for another, and he gets in deeper and deeper. The next thing you know he's running to the bulls or the DA's office. It's the same with working people. Those are the ones you got to use force with, so I don't bother with them.

At one time I had around 150 regular customers. I got rid of the ones that were headaches and kept the ones that were no trouble—bookmakers, numbers runners, guys in illegal stuff, maybe some saloonkeepers, that line of people. Truthfully, I did not use any muscle with them. That's why I was known as the best shylocker all around. I was smart. What's the difference if I let somebody skip for a week? My other money is circulating, so I am not going to worry about this par-ticular person that has gotten in trouble, maybe a bookmaker who has got hit hard. Who can say what it might be? I figured the wisest thing to do was to make it for his convenience. That is the way I operated.

Let's say a guy needs $1,000, but he can't make the $100 a week. If he has a good record, I'll cut him down to say, $80, and I won't charge him any extra. Naturally, as soon as the money comes in, I would give it to somebody else. It circulates. By the time it circulates around, I never could figure out what the percentage is I'm getting on it. It would be impossible for me. You would need an accountant to figure that out.

A steady clientele, without the necessity of beating up any debtors, was especially desirable because, as Valachi notes, "You find, as you go along, that most of these people get in the habit of reborrowing before they pay up." For a loan shark like Valachi this was where the real windfall lay—in a reloan, or, as it is called in shylocking circles, the "sweet" loan. When a borrower already in debt wants more money, the loan shark simply deducts what is still owed him, but charges for the entire amount that has been requested. This has the practical effect of doubling the interest rate:

I'll give you an example. There is this guy Hugo who is a bookmaker. One day he's hit hard, and he's still into me for $300 on a $500 loan. Now I go over to see him to make my regular collection. He is paying me $50 a week, but now he's in trouble, and he wants a reloan of $500. He already owes me $300, so all I have to do is hand him $200 more. But I charge him $100 as though he had just borrowed it all because that's the vigorish on $500. This is why the cream is in the reloaning. All I'm really doing is giving him $200 in cash to get back $100. Giving him a break is wise. The idea is not to bust a guy's head. Anybody can do that. The idea is to keep the money moving all the time.

Operating on his own in this fashion, Valachi soon had about $10,000 "out on the street," which was bringing in an average

$1,500 a week. "But I got," he is quick to add, "a lot of expenses, too. It seems like somebody is getting married every other day, and that alone is costing me $50 to $100 for the envelope, as I don't want to look cheap." Then his shylocking expanded considerably with the acquisition of a partner, John (Johnny Roberts) Robilotto, who financed him. In return Valachi was responsible for developing new customers and managing the racket:

Johnny was the best in the land. He was heavyset, about five foot eight, and believe me, he was the kind of guy who couldn't say no. He was partners with Tony Bender in these clubs. Some of them were the Hollywood, the Village Inn, the 19th Hole, and the Black Cat. Whatever Johnny used to do, it seemed Tony was partners with him. Don't ask me why a nice guy like Johnny would want to get mixed up with Tony. Who can explain those things? Certainly you ain't going to ask him.

Johnny wasn't a member then. He was with Tony Bender, but Tony was never able to get him in because Johnny had a brother who was a cop. In other words, even Tony didn't have enough weight to get him past that. Later, a lot later, Johnny started hanging around with Albert Anastasia, and Albert got him in his Family. You understand what I mean when I say "in." He got him in the organization, this Cosa Nostra. It happened all of a sudden. How or why I don't know except, as I will explain when the right moment comes, you never fucked around with Albert Anastasia. Naturally I didn't question it. I was just happy to hear Johnny was a member.

Valachi first met Robilotto in one of his clubs. He had heard that Robilotto was running a large still near the Hudson River waterfront. "Of course," he notes, "Prohibition is over for a long time, but this don't mean there ain't a market for alcohol." Valachi knew of some potential customers in the Bronx for illegal alcohol, and he hoped to earn a commission by bringing them and Robilotto together. "I thought it would be a good deal for Johnny," he said, "as these people will take all his output, and he don't have to run around selling it here and there." Robilotto, however, told him that he had as much business as he could handle. Valachi accepted this at face value and let the matter drop. A few days later he was outraged by a visit from Bobby Doyle. "I hear you got some people for Johnny's alcohol," Doyle said. "I spoke to Tony about it, and it's all set."

For Valachi it was another example of Doyle's venal ways. "See what a dog he was!" he says. "He's got to have his hand in everything I'm doing. Well, that means I got to split the commission—which is about $500 a week—with Bobby." Valachi found Robilotto apologetic about the whole thing. "I couldn't make a deal with you," he said, "without Tony's okay."

"Johnny," Valachi said. "I understand. It ain't your fault."

The two men subsequently became friendly. Then one night Valachi was in the Hollywood when a shylock customer asked for a loan of $2,000. Valachi was unable to oblige him. "All my money," he explains, "was out on the streets." Robilotto, however, overheard the conversation and advanced the money to Valachi. After he had paid it back, Robilotto became increasingly interested in learning how Valachi operated. Finally he said, "I like your style of doing things. Maybe I'll go partners with you."

Shortly afterward, Robilotto gave him $20,000:

He tells me to take the money and put it out, and we will share the income. "Tie it up," he says, "so I can't get at it." Believe me, he's telling the truth. He is a wild man with money. I remember meeting him at this crap game on the East Side somewhere, I think on 14th Street, and I never saw anybody lose so fast. Before I can turn around, he's out $15,000. I lend him a couple thousand, all I got in my pocket, and that goes too. He borrows $5,000 from the house, as he is known, and now he is down to his last $600 when this kid Joey, who works for me in the numbers and the shylocking, comes in looking for me as I had left word where I was. This Joe is lucky, he gets hot and he can make ten straight numbers, so I say to Johnny, "Let him shoot for you."

It's like a miracle. Joey's first point is four, and that's a rough number. He makes it the first time out. Then he has two sevens in a row and wins again. Joey's next point is five, and Johnny goes the opposite way, meaning he bets on nine all around the table, and Joey throws two or three, I forget how many, nines, and now Johnny goes to five, and Joey throws a five, and the place is a madhouse. It keeps up that way, except we got to sweat a little as Joey's point is ten and Johnny has got it all on ten, and Joey shoots maybe six or seven times before he makes it. To make it short, Johnny gets ahead a few thousand, I got my money back, and I pull him from the table, and he gives a thousand to the kid. Then I hear that the game is sore at me because I made Johnny quit, and they are thinking of barring me. Who the hell cared? I didn't like to shoot crap that much anyway.

Well, with the money from Johnny Roberts, the shylocking is doing pretty good. I'd say we had around $60,000 out in the street, and naturally Tony Bender knows about it. Tony is losing heavy at the track, and Johnny gives me a hint one day, saying Tony was asking all kinds of questions about our shylocking together, meaning Tony wants a piece. Being it's Tony, I can figure that the piece he wants is a third of the action. Johnny is under Tony's thumb at the time, and he can't do nothing, but I decide I ain't going to stand for it. I already was shaken down once by Bobby Doyle with Johnny's alcohol output, and that is the last time it's going to happen. Pretty soon the word will get around that if you want to shake somebody down, shake down Joe Cago, he ain't so tough. So I tell Johnny, "Look, if Tony is asking you questions and you are thinking of giving him an interest in the business, you talk for yourself, because I won't give up any of my share."

The next thing I know is that Tony Bender sends for me, and we meet at Duke's restaurant in Jersey. It is in Fort Lee or Cliffside, I forget which, as those towns run together and you never know where you are. Anyway, all the boys living over there hang out at Duke's.* Tony tells me that the partnership between Johnny Roberts and me was going to be broken up and to collect all the money. He says he is sorry, but he needs cash fast on account of the horses.

I tell Tony that the money is all out, everyone is keeping up their payments, and I ain't going to push them. After all, I got the business to think about. I say I will figure out Johnny's share and borrow it from another shylock, but I will deduct the interest, which is 10 percent.

"That will eat up all what he has earned," Tony says.

"Listen," I say, "none of this was my idea.

You want the money, that's the way it must be."

He says, "Okay."

So this is how I broke up the partnership. Maybe I was wrong. But in plain English I would rather break it up than give Tony Bender any money or let him in the business, and when all is said and done, I still got around $30,000 of my own circulating now, and I am satisfied.

_____

*Duke's restaurant was in Cliffside Park, New Jersey. Until it closed in 1950, it was a favorite gathering place for Cosa Nostra chieftains who had taken to suburban life, among them Willie Moretti, Vito Genovese, Albert Anastasia, and Joe Adonis.

Valachi's loan-sharking methods inadvertently led him into an area which today is all the rage in Cosa Nostra—the penetration of legitimate enterprises. Instead of trying to squeeze money out of a client who had defaulted on a loan after suffering substantial gambling losses, Valachi accepted his invitation to take over a half interest in a restaurant in upper Manhattan called the Paradise. As usual, he did not make the move precipitously. "I told him I would think about it," he notes. "This guy—his name was Eddie—was no dummy, and I wanted to make sure there wasn't a catch to it." First he observed how much business the restaurant was doing over a period of days and then spent some time quizzing one of the bartenders. After some haggling over the worth of the partnership Valachi finally agreed to $9,000, deducted the $3,500 owed him, and gave Eddie the rest in cash. "So now I got my own joint, or half of it anyway," he told me, "and I make it my hangout. I do my shylocking there and I bring in a new chef, so all the boys will drop in. I'd say my end was about $800 a month."

Valachi's participation in the restaurant could not be officially noted since his police record made him ineligible for a state liquor license. To protect himself, he had a private agreement drawn up with his partner's wife as a witness. He was now anxious, however, to have a tax cover to explain his livelihood. The problem was solved when a similar set of circumstances landed him in dress manufacturing:

There was this Matty who had a dress factory in the Bronx at 595 Prospect Avenue—that's why it was called the Prospect Dress and Negligee Company—and he was one of my best customers. With reloans and whatnot he was into me for quite a few thousand but was paying up regularly every Friday. All of a sudden he asked if I could let him pass for a couple of weeks. Well, when these two weeks are up, he asks me to let him ride for another week. Now I go up to see him and I ask, "Matty, what's the matter?" and he explains that he is having a hard time getting business from the jobbers because his machines—the kind that make buttonholes and everything—are wearing out. I look over the factory, and it looks pretty good to me. Matty is no fool, and he can see what's on my mind, and he says why don't I go in with him. So I talk to his main jobber, and he tells me that if Matty gets in better equipment, he will give him all the work Matty can handle. He says he will give him business where the styles don't change every time you turn around.

Now the next thing I got to worry about is the union, so I go downtown to the garment district to see Jimmy Doyle, right name Plumeri, or one of the Dio [Dioguardi] brothers, Johnny or Tommy. I forget who, I think it was Jimmy, but it don't make no difference as the Dios are his nephews, and when you talk to one, you are talking to all of them. They are in Tommy Brown's Family, and they are supposed to straighten out any trouble with the union, I think it was Local 25. I tell Jimmy Doyle what's what and how I'm figuring to go into the dress business, but I don't want no unions around, especially because the jobber I'm depending on don't handle union work. Jimmy says to me, "Don't worry. As long as you are in the Bronx, it's okay. It's no sweat. We can handle it. But stay out of Manhattan."

Well, that's all there was to it. I go back and tell the jobber that I'm going to be Matty's partner and the union business is all taken care of. He says, "What did you do?" I just mentioned to him the names of the people I had been talking to downtown, and I must say he was impressed. I'd say I was in the dress business for about twelve years, and we only had a couple of complaints. If any union organizer came around, all I had to do was call up John Dio or Tommy Dio and all my troubles were over. One day I go into the shop right after a union guy had come in and pulled what you call the switch. In other words, he shut the place down. I said to Matty, "Stupid, don't let anybody do that. Didn't I tell you I got everything fixed?" I would have run the guy right out, but Matty was a legitimate man. He got scared to death and folded up. This union guy told Matty that he was going to report us, so I got on the phone and called Johnny Dio, and that was the end of it.

A long time after that two union organizers show up again, and this time I am there in the place. They want to know if our girls are in the union. Most of the girls are Puerto Rican, and I say to these guys, "Are you crazy? They don't even know how to speak English." Now I just don't like unions. To me a union guy is a pimp. I am sorry, but that is my attitude. They want to know how much we are paying the girls, and I find out it's more than union scale anyway, so I'm twice as sure not to let them force me into the union. They tell me they're going to have to make a complaint, and I get tired of all this talk. Matty has a permit for a pistol, and I go over and get it out of the drawer. You should've seen those guys run. Johnny Dio calls me in about an hour, and he is laughing as he has heard what happened. "That's the way to handle them," he says. "They said there was a wild man up there and they were lucky to get out alive. Don't worry, they won't be back."

Well, counting what Matty owed me when I went in with him, I'd say I laid out around $15,000 to go into the business. We had to buy all kinds of special machines for things I never heard of, like for bottom stitching. These specials cost money. Matty, being a good machineman, bought a lot of secondhand stuff and put them in good working order. Out of my investment I was getting back maybe a couple of hundred a week, which is okay because Matty is running the place and I got something for the tax people if they start nosing around.

I don't butt in, except once when I noticed there was something wrong with the profits as all of a sudden my share dropped to $100 a week. I asked the accountant what was going on, but I guess Matty had warned him not to tell me the truth—that he, Matty, was losing heavy on the numbers. I find out when Matty comes to me and says he has hit a number for $2,400 and they won't pay off. Naturally, I went to the bank, and believe me, I collected the money for Matty. The numbers guy tells me that Matty has been playing all this time, and that's where the money was going. I take out what's owed me and give the rest to Matty. Then I give him a slap in the face that knocks him back about ten feet. "So that's what you're doing with the money," I say. "You're playing like a sucker, and when you finally win, you don't even get paid." I tell him I'm in the numbers and to make sure he don't play anymore. I say that he was lucky to win once, as most of the time the numbers are fixed. That stopped him, and later his wife—she was a fine lady—comes and thanks me because she said she was going out of her mind with Matty's gambling.

Well, I got the restaurant and the factory. I'm still in the numbers with that dog Bobby Doyle, and I got all my shylocking. So I make up my mind to take it easy for a while and to stay away from racket guys. In the old days I was taught if I got to belong to a mob, I got to be a man and not a lob [flunky]. Now I got Tony Bender over my head. I'm getting nothing from him, and if he thinks he's going to get anything from me, it will be over my dead body. I'd rather go to work if I got to be a lob for him. In other words, I am starting to be a boss hater. Who can blame me?

By now we got the kid—his name is Donald—and naturally Mildred wants to stay home at night with him. But I can't do it. Thinking about all these things is driving me nuts, and I got to move around. So I start going out with some girls just to relax, going here and there, as sitting at home every night is too hard on me. There were six of these girls all living together in East Harlem. I met the first one, Jeannie, through one of the boys. There was also Laura, Rose, Helen, Louise, and I forget the last one, but I liked Laura the best, and as time went by, I set her up in her own place. They were around nineteen, and I would give them about $40 to $50 to buy a dress when I went out with them because I would be ashamed to be with them if they don't look good. I forgot to say that first I sent them to this doctor for an examination. It was a good thing. Two of them had to be taken care of. "I can't let them go on the street like that," the doctor said. He charged me $80.

8

The late 1930s were a generally unsettling period for organized crime. Luciano and Genovese were not the only ones either to fall or flee as a result of the Dewey crusade against racketeering. Next on the list was the most feared Jewish mobster of the day, Louis (Lepke) Buchalter. His ultimate fate—the electric chair—is especially noteworthy since no Cosa Nostra chieftain of similar status has ever been legally executed. Indeed, once in power, they rarely saw the inside of a cell.

Buchalter got his nickname from a doting mother who delighted in calling him Lepkela, or Little Louis. Slender, reticent, and sad-eyed, he looked more like a shoe clerk than the mastermind of a huge narcotics ring, as well as the absolute ruler of labor and management extortion in trucking, in restaurants, in movie theaters, and in the baking, garment, and fur industries.

But as big as he was, Buchalter simply did not have the resources and the mystique of a Cosa Nostra to enable him to survive. By 1937, wanted for murder and indicted on narcotics and extortion charges, he decided to hide out while methodically directing the demise of various people who could testify against him. Then one of his gunmen made the mistake of shooting down a luckless citizen who happened to resemble a potential Dewey witness. The ensuing public uproar triggered a massive manhunt for Buchalter dead or alive. The strategy used in trying to flush him out was so to harass his Cosa Nostra colleagues that they would begin having second thoughts about his continued freedom. It worked. Not even Buchalter's closest ally in the Cosa Nostra, Albert Anastasia, was able to intervene in his behalf. Finally, after two years as a fugitive, Buchalter's underworld brethren sold him on the idea that a deal had been worked out with federal authorities. If he surrendered on the narcotics charge, he would be allowed to serve his sentence before New York State could get a crack at him for murder. The plan had an undeniable appeal; who knew what misfortune might befall witnesses against him while he was in prison? But Buchalter soon found he had been duped. Within seventeen months he was on trial for his life and, doomed by the testimony of one of his own men, received the death penalty.

On the heels of this, the revelations of Abe (Kid Twist) Reles kept public attention focused on organized crime. Reles, a Buchalter crony, belonged to a band of Brooklyn mobsters popularly called Murder Incorporated, which had strong ties with several Cosa Nostra figures, most notably Anastasia. Information from Reles led to convictions in a half dozen previously unsolved gangland slayings. Murder Incorporated has since been portrayed as a sort of specialty house which handled contracts to kill for the entire U.S. underworld. This, according to Valachi, was not so at least as far as official Cosa Nostra executions were concerned. In such matters, he says, it has always relied on its own membership.

The career of informer Reles was cut short on November 12, 1941, following his fatal plunge while in protective custody from a sixth-story window at the Half Moon Hotel in Coney Island. The chief beneficiary of his death, curiously enough, was Albert Anastasia; as then Brooklyn District Attorney William O'Dwyer noted, a "perfect" murder case against him "went out the window with Reles." The debate over how Reles happened to fall with six policemen guarding him raged on for years. For Valachi, however, it was not much of a mystery. "I never met anybody," he says, "who thought Abe went out that window because he wanted to."

Thus, quite aside from his disenchantment with "mob life" brought on by his running feud with Tony Bender, it was an ideal time for Valachi to lie low. And comfortably off with his income from the numbers, loan-sharking, his dress factory and restaurant, he decided to take a fling at racing horses. It is probably the only thing he ever did for the pure pleasure of it. During my interviews with Valachi, he could speak with utter dispassion about his involvement in a Cosa Nostra contract, his acquisition of a mistress, even his "beefs" with Bender. When he spoke about his horses, he became another person, jumping up from his chair and excitedly pacing his cell, acting out each story, for once eager to relive his past.

Valachi's underworld ethics crumbled the first time one of his horses, Knight's Duchess, seemed likely to win a race. The only other horse entered that concerned him was owned by a friend named Joseph Bruno, and on the eve of the race, at Rockingham

Park in New Hampshire, Bruno approached him with a proposition. If Valachi agreed to let Bruno's horse High Caste win, Bruno
would return the favor the next time they were in competition.
That way, Bruno argued, they could realize a small fortune since
both horses, while in top form, were long shots as far as the betting public was concerned. Valachi indignantly rejected the idea. "I
told him," he says, "I didn't care nothing about the money. All I
wanted to do was win my first race. I was so nervous I was already
sure I wasn't going to sleep that night. Only a trainer or another
owner will understand my feelings."

The details of the race are forever etched in his mind:

Who can forget something like that? First I got to tell you about this
Knight's Duchess. I just loved the name. She was from Ireland, and
she was what you call a bay filly. She was by Knight of the Garter out
of Diomedia. You see, real class. She was listed as a two-year-old when
I bought her, but she was really one and a half. I'll explain why. Any
horse no matter what month he is born automatically becomes one year
old the next January. It's the rule. Being as Knight's Duchess is born
in Ireland, she runs the wrong way. Over there they make the horses go
clockwise, don't ask me why, so naturally we got to teach her the right
way. Now this takes time, but she's ready to run as a three-year-old. As
she is being trained, we find out some things about her. We find out
she is best at a distance than in a sprint. One morning we clock her at
one forty-three for a mile, and she is faster during the last sixteenth
than she was at the first. The trainer tells me she is ready and we got
to run her at Empire State racetrack. That's when I find out the second
thing about her. "What's so important about Empire?" I ask, and he
explains that Empire has a very hard track and Knight's Duchess will
do well there because she has small hooves. In other words, if a track

is soft, she is going to sink in and get tired. Gee, I think to myself, how much do you have to know to race horses?

Well, before we can do anything, they sell Empire State and make it into a track for trotters. That's why we ship her to Rockingham. It's a hard track—not as hard as Empire but hard enough for her to bounce off as she is running. Now Knight's Duchess is a long shot because the first time she races at Rockingham she finishes out of the money. I wasn't there, but the trainer telephones me and says he don't like the way the jockey handled her. To tell the truth, he thinks the boy pulled the horse. So I go up there myself, and I hire Teddy Atkinson—you must have heard of him—to ride her. I want the right boy to be on her this time. I know she is ready. I got a thousand on her to win and a thousand to show. She starts off at 40 to 1, and at post time she is 11 to 1. You never can be sure, but I think all I got to worry about is this High Caste. Knight's Duchess is in post position number seven, and High Caste is in number five. I thought I'd have a heart attack.

I will call the race for you. They are in the gate, and they're off. This horse—I forget the name, it was something like Brave Action—breaks in front, another horse is second, High Caste is running third, and Knight's Duchess is fourth. As they go around the first turn, Brave Action is now out in front by four lengths, this other horse is running second by four lengths, High Caste is third by three lengths, and Knight's Duchess is still fourth. Down the backstretch Brave Action is in front by half a length, High Caste is catching the second horse, and Knight's Duchess is four lengths off the pace. They are around the stretch turn, and High Caste takes the lead, the other horses are out of it, and here comes Knight's Duchess.

In the stretch it is High Caste by a head; Knight's Duchess is second and coming on. Now she is starting to pass High Caste. All of a sudden the boy on High Caste raps Knight's Duchess in the mouth with his whip. I see it, and I'm up on my feet screaming and yelling. Everybody sees it, and they're yelling, too. It happened right in front of the grandstand. Knight's Duchess comes on again, but it ain't no use, and she finished second. They put up the winning number, and it's number five, High Caste. I'm already thinking what's going to happen to that jockey, even if he is Joe Bruno's boy, when Teddy Atkinson goes to the stewards and claims a foul. I hold my breath these few minutes, and then they change the number, and they put up number seven, which is Knight's Duchess, and I am a winner.

Boy, was I happy! I don't care about the money, but it ain't so bad. I am ahead $10,400 on my win bet, and I also collect for show. I think it was around $2,000. I don't bet at the track, as that would knock down the price. I bet with bookmakers back in New York. Now I want to explain something. When I placed my bets, I insured them. The insurance is 10 percent of what you bet. If you don't do this, a bookmaker would only pay you up to 20 to 1 even if the horse come in at more than that. By insuring your bet, you collect whatever the horse pays at the track. Of course, Knight's Duchess closed below twenty to one, but only at the last minute. In other words, it is wise not to take any chances when you have a long shot.

After the race the first thing I do is to give Teddy Atkinson $200 or $300, plus the fee for riding the horse, which is $25. I give the trainer $500, and I tell him to keep the purse, I think $800, as there are a lot of expenses in running horses for shoes and medicine and whatnot. I don't forget the exercise boy and the stable hands. I don't give them much, just enough so that they know I am thinking of them. It is important to have everybody on your side. These people can do you harm. If they want, they can give you all kinds of phony reasons why your horse is losing. I remember a lady had this fine horse, and he was always losing. Why? Because the trainer was having the jockey pull the horse, so he would get the lady thinking that the horse ain't no good and he could buy him cheap. "Well," the trainer would say to this lady, "the horse just didn't seem to want to run today. I don't know what's wrong with him." It is them kind that can give the horse business a bad name. You got to watch them all the time.

Now, before I forget, here is the best part of my story about Knight's Duchess. I am back home, and the guys in this horse room tell me that Bobby Doyle bet a bundle on High Caste, and as High Caste was disqualified he blew his top and starts screaming, "Joe must have fixed the race!" I got a good laugh out of it. That's the way a dog like him thinks.

Until 1937, as far as Valachi was concerned, thoroughbred racing's sole function was to determine the winning policy number each day. Then he took his wife, Mildred, to Miami for a winter vacation, and in the company of The Gap and various other Cosa Nostra members he began frequenting the track. At first he was not especially taken by it. "With the tips all my friends gave me," he says, "I was lucky I was only betting tens and twenties. I got nothing but losers."

A chance encounter, according to Valachi, changed everything. One afternoon Mildred bet $2 on a horse listed at 99 to 1 because she liked his name. The Gap and Valachi ridiculed her choice, but an elderly man nearby told her, "Don't listen to them, lady, I think you're right." When the horse won, Valachi decided the stranger was worth cultivating:

This was how I learned to play horses. I went to him and said, "Thanks for the tip you gave my wife. It's the first good one we got since I been here." He said no thanks are needed; he was glad to do Mildred a favor as she looked so nervous when The Gap and me said her horse ain't got a hope.

I asked him, "Was the race fixed?" and he looks at me kind of funny and says, "How long have you been coming to the track?" I tell him the truth, which is not long. He laughs, and he says, "Well, that is the first thing you must learn. Even when the fix is in, nothing is ever certain." Then he tells me about this guy that fixed a race to make a killing like we all dream of. The guy has eight or nine horses entered in the same race with different people fronting for him. It is a race for nonwinners that they have every now and then to give a trainer a chance to come in with a horse that ain't won yet during the year. All the guy's horses fit in this class, and that's how he can get them together in one race. Naturally there are some other nonwinners entered, but this guy pays off the trainers, and they are scratched. So now he is all set. He can't lose as he has given instructions about which horse is to win. He spreads his bets around, and he stands to collect a couple of hundred thousand. Well, the horse that is supposed to win has a fine lead coming down the stretch. He is all by himself, and nobody could catch him even if they wanted to. Then what happened? The horse stumbles and breaks a leg, and that is the end of the fix.

"Gee," I say, "I see what you mean, but how did you know the horse my wife bet on was going to win?"

So he says that's what he is trying to tell me. He didn't *know* the horse was going to win. Every horse in a race has a chance, he says, and I should never forget it. That's the beauty of the sport. But he says he had a pretty good idea about this horse. He knew they have been bringing the horse along, and one of the stableboys told him the true time of his last workout, which was three seconds faster than what they gave out to the newspaper handicappers. You see, it is to their advantage to hide a horse's ability whenever they can.

"Oh," I say, "the idea is to get connected with the trainers and them others working around the barns."

"Yes," he says, "if you are going to be serious about racing. That way you have an edge. It's all you can ask for. You listen to all these tipsters, and you'll just lose your shirt."

Well, to make it short, he was a nice old man—that's why I am leaving his name out of this—and he has been following the horses all his life. I stuck with him the rest of the time I was in Florida, and he introduced me to a lot of racing people. When the season come to New York, I started going to the track, mostly with Johnny Roberts, as that was the time we were getting friendly, and I looked up some of the guys I met through the old man. I also get to be a pal of this handicapper who has good information and is what you call a fine judge of horse-flesh. With one thing and another, I'd say I am ahead between $200 and $300 a week, and I am having a lot of fun. I don't win every week, but it averages out to that. I'm not foolish. I don't bet every race, only when I got a horse that looks good, and when I bet, I bet two ways— win and show—to cover myself.

Now the word gets around among the boys that Joe Cago—meaning me—is doing okay at the track, and one day Tony Bender calls me and says to meet him that night at Duke's in New Jersey. I ain't in the habit of going over to Duke's except when I'm on the carpet for something, so as I walk in, one of the guys that hangs out there sees me and says, "Oh, God! Are you on the carpet again?"

"Nah," I say. "I got an appointment with Tony."

It don't take no genius to figure what *Mr.* Bender wants. Tony was a real sucker with horses. If my grandmother told Tony about some horse in the sixth race, the only sure thing is that Tony would bet on the horse. A lot of mob guys are like that. They are supposed to be so tough and smart, and they act like the biggest squares in the world when they get a tip—it don't matter from where—on a race.

So after talking about this and that, Tony says, "Joe, I need a winner." Now this is before he made me and Johnny Roberts split up in the shylocking business, or I wouldn't do what I done, which was to give Tony a horse. This kid, I think he was with the George D. Widener stable, told me about a horse that was ready, and he is running the next day. I still remember the horse's name. It was Harvard Square.

Tony has the form sheets with him, and right away he looks up the race and says, "Hey, he is up against a horse that ain't lost this year."

I say, "I ain't going to argue with you. This kid told me the other horse will get beat tomorrow, and the kid knows what he is talking about. Do what you want."

Well, Tony bets Harvard Square, and Harvard Square comes on in the stretch and pays $12 and change to win. Tony is a very heavy bettor, so this means he wins a lot. I tell the story because it is the only time I ever got satisfaction from Tony Bender. He is so happy about that he forgets himself. He went around telling everybody, "Boy, Joe can really pick them!"

Naturally I am sorry I ever made him win. Tony ain't with us on earth no more, and when the right time comes, I will explain how he got his, not that I'm losing any sleep over it.

Valachi's fascination with racing inevitably led to the purchase of his first horse. At the end of the 1940 racing season in New York a contact informed him that the owner-trainer of several horses, unable to pay his feed bill at the track, was willing to part with a five-year-old gelding named Armagnac, provided he was kept on as the trainer. Valachi knew about Armagnac, having bet on him and won a few weeks earlier, but he was wary about the conditions of the sale. "What kind of a trainer can he be," he asked, "if he is broke with a fine horse like that?"

"It's just one of those things," he was told. "The man is a fine trainer, but he's a gambling fool. He bets every race, even when he knows better. You'll be all right so long as you don't let him tout you."

The temptation to have his own horse overcame whatever doubts Valachi entertained. His police record was no problem. With Armagnac, as with his other horses, he simply had various people, including his wife, front for him as the owner of record:

Well, I got a horse, but I can't run him right away. The trainer tells me that Armagnac is tired and should rest over the winter. I am disappointed, but the horse must come before my feelings. We figure out what the bill will be for putting him on the farm—which was around a couple of thousand—and I can't wait for the next season.

When the time comes, Mildred is so excited she can hardly talk. I try to explain that the first race Armagnac will run in is just a test. It ain't meant for him to win. She don't understand. She says to me, "Why are you going to race him if you don't think he can win?"

I try to explain, but she still doesn't get it, and nothing will do but that she must come and see him. Then when we get to the track, I see that Armagnac is in what they call the field. I'll explain what this is. If there are more than twelve horses in a race, they put these extra horses in the field. There could be three or four horses in the field, they are usually long shots, and if you bet the field and one of them wins, it don't make any difference which one, you win. So I find that one of the horses in the field with Armagnac has got a chance, and I tell Mildred to go ahead and bet on Armagnac across the board.

They're off, and at the wire they are all bunched up and Mildred could not tell where Armagnac is. He ran a nice race and I think he finished sixth, but this other horse in the field comes in second, so Mildred collects place and show money. "You see," she said to me, "you must have faith." To this day she thinks it was Armagnac.

Initially, however, Valachi enjoyed considerably less success as an owner than he did as an amateur handicapper. The next time Armagnac was scheduled to run, the morning line, or advance odds, on him was 50 to 1. "Now he is very sharp," Valachi recalls, "and there ain't another horse in his class in the race. I am going to put $2,000 on him to win, and even if the price drops, I will clean up." But when Valachi arrived at the track, he learned that during the night Armagnac had badly bruised a foreleg against the side of his stall and would be out most of the season. Valachi took the news philosophically. "What could I do?" he says. "After all, the poor horse didn't kick himself on purpose. I advised myself to be patient, as there is always another day."

By now he also had Knight's Duchess. But she, as Valachi puts it, was still being trained to "run the right way." He then acquired a two-year-old colt which he called Walter Raleigh after his sire, Sir Walter Raleigh. "I figured I would just Americanize the name," Valachi says, "but I had to do a lot more than that. Walter was a bad boy. He was a lover. Instead of racing, he wants to jump every filly in sight, including Knight's Duchess. So he has to have an operation. In other words, his balls got to come off, and he will be a gelding. You see what you must go through having horses?"

But Valachi the wide-eyed owner was one thing. He was on considerably firmer ground when, as happened periodically, someone tried to cross him at the track. A memorable attempt took place shortly after Knight's Duchess had won in such dra-matic fashion at Rockingham Park. He says it marked the first—and only—time he deliberately had one of his horses lose:

I don't want to do it, but I got no choice. If I let this guy clip me, the word will travel, and everybody will be trying something against me. That is the way mob life is. Now the guy's name was Charley. He is in the furniture business, and all the guys with Tony Bender are buying furniture from him, as the price is right. Then he gets in some kind of trouble and Tony lends him around $15,000. Naturally he has to put up his inventory as security.

Around this time he comes to me and says he has got a lot of people who will bet for him if he has a good tip on a horse. He gets half the winnings. He says to let him know when Knight's Duchess is ready. He will recommend her, and if she wins, he will split with me.

Well, I like the idea. The trainer already has Knight's Duchess entered in a race which looks good for her. So I tell Charley, and he says, "Let's go." I take my girl, Laura, and I get this other girl, Helen, for him, and it looks like we will have a swell time and make some money, too.

The morning of the race I smell something fishy. We are at the hotel, and Helen comes into my room. I ask, "Where's Charley?" and she says he has gone to the lobby to make some calls. I don't understand why he ain't calling from his room, so I send Helen down to see if she can learn what this is all about. She comes back and says he is making one call after another advising people to bet on Knight's Duchess.

I still don't know what's going on, but I find out. We are on the way to the track, and Charley keeps asking, "Are you sure she's going to win?" I tell him that Knight's Duchess is a cinch unless—God forbid—she breaks a leg. Then all of a sudden at the track he says, "Joe, I ain't going to make any calls. I just don't feel lucky today."

That's all I have to hear. He is planning to cut me out of my end of the bets he told his people to make. I run to the trainer and tell him

that we got to teach this guy a lesson. The trainer says that he will take
care of everything with the jockey and not to worry about a thing. Now
I figure one of the people Charley called is Tony Bender. There is no
need getting Tony steamed up at me for nothing, so I call him and warn
him not to bet on Knight's Duchess.

After that I sit back and wait. I can hardly keep from laughing as I
watch Charley's face. Knight's Duchess ain't never in the race, and I
think she finished last. The sweat is pouring off Charley, and he jumps
up and runs off without a word. I follow him and see he is on the phone
again, and when he ain't on the phone, he is drinking pretty good. I got
an idea what's coming next. Sure enough, when he is back, his face is
all red, and he is shouting. "Did you make a call to Tony Bender?"

"Gee," I say, "that's right. I guess I forgot to tell you."

Now I let him have it. I tell him everything I done, and I thought he
was going to make a move for me. I wish he had. I would have busted
his head in. But he folds up and starts moaning, "I laid out all the furniture money...I told all my people to bet...what'll I do?"

I said, "Charley, you wanted to fuck. You got fucked."

Although Valachi's horses, notably Knight's Duchess, went on
to win a number of races, his interest in them was diverted because
of pressing economic problems brought on by the war. "I mean
the war when the Japs bombed us," he was careful to point out to
me, "not some trouble in the Cosa Nostra."

Its most immediate effect was to end Valachi's resolve to stay
away from "mobs guys" as much as he could. He had by then sold
his share of the Paradise Restaurant, at 110th Street and Eighth
Avenue, since "the neighborhood was changing; the colored were
moving in, and they don't eat our kind of food." This left Valachi
with his numbers, his loan-shark racket, and his dress factory. Of

these, only the factory, which had shifted production from its normal line to fill military orders, profited from the war. "Now," he says, "the numbers and the shylocking go dead on me. There is plenty of money around and plenty of jobs, so who needs to borrow? It's worse with the numbers. I'll tell you something about them. The numbers are good only when times are bad. It's poor people that play the numbers, and if you want the truth, most of them play because they are desperate for money and they don't have no other way to get it."

A traditionalist at heart, Valachi was reluctant to give up such an old standby as the policy game. But as his bank's daily play dwindled, he also ran into a streak of "extra bad" hits that forced him to dig into his capital to pay off. If this was not dispiriting enough, many of his best runners, vital to the success of a numbers operation, began drifting into war work. "When I lost my help," he recalls, "I quit. I am out $23,000 anyway. I am sick and tired of the numbers and being shaken down by the cops, and I got to find something else."

By now Valachi's two good friends Frank Livorsi and Dominick (The Gap) Petrilli were deeply involved in bringing morphine in from Mexico for conversion into heroin. They invited Valachi to join them, but after thinking it over, he declined. "At the time I don't know too much about junk," he says, "and I was suspicious of it." Valachi's instincts were correct. Within a year Livorsi and Petrilli, among others, were rounded up and sent to prison in the first major case built by the Bureau of Narcotics against the Cosa Nostra.

His problem, however, of what to do was solved soon enough. For the Cosa Nostra the war, like everything else, was simply a situation to be exploited. The only question was how. As a host of shortages developed in the economy, necessitating price rationing and price control, the answer was evident—the black market. It was not perhaps quite the bonanza that Prohibition had provided, but the same classic mix was present for the organized underworld to fatten on: the illegal sale of a commodity—be it meat, sugar, gas, etc.—to a public that by and large was a willing, indeed anxious, accomplice.

This combination assured not only huge, tax-free profits, but also little risk. Valachi specialized in gas ration stamps, and from mid-1942 until 1945 he made about $200,000—"I wasn't so big," he modestly notes—without any interference from the law.

Valachi confesses that he did not immediately realize the potential of the stamp racket. "I thought it was penny-ante stuff," he says. "Then I find out how them pennies can mount up." He first got involved at the behest of the owner of the garage where he kept his car—"Joe, you got connections; I know you can get me stamps." Valachi finally agreed to see what he could do. Since there were various types of gas stamps, he had the garage man list the kind he wanted, the number of gallons, and the price he was willing to pay. A few nights later he bumped into "one of the boys who was in stamps"—another member of the Cosa Nostra named Frank Luciano—and asked what the order would cost. "Well," Valachi recalls, "I look at the price the guy at the garage will pay and what Frank will charge me, and I see I stand to earn $189. Now this was a small deal, I think only 10,000 gallons, but what did I have to do? One guy gives me the money. The other guy gives me the stamps. Boy, I say to myself, this is the business for me."

Valachi's next transaction with Luciano, for 100,000 gallons, netted him $1,700. "I don't remember the price," he says, "just the profit." Soon afterward he eagerly accepted Luciano's suggestion that they form a partnership. Essentially Valachi became a wholesaler in a carefully structured operation. It was extremely important, in his words, to be "mobbed up." This guaranteed that all the stamps he handled would be genuine. "That was the catch in the business," he said. "It was dangerous fooling around with fake stamps. They could trace them too easy. They put them under these special lights which would bring out the phony ones. It was like trying to pass counterfeit money. There ain't any percentage in it."

Legitimate gas ration stamps were obtained by breaking into local boards of the Office of Price Administration. This was relatively hazardous, and according to Valachi, it was generally left to independent gangs, which, lacking the organization to distribute the stamps, then sold them in bulk lots to Cosa Nostra dealers. In time another source was developed. "With all the burglarizing," Valachi notes, "a lot of OPA offices stuck the stamps in banks overnight. So the next thing is certain OPA people themselves started sneaking them out and selling them."

Since the average profit on "black gasoline," as it was called, ranged between three and five cents a gallon, the volume had to be large enough to make it worthwhile to the organized underworld. It was. OPA records at the time showed that at least 2,500,000 gallons were being diverted for illegal use *every day* throughout the war. It was next to impossible for the OPA to cope with the alliance formed by the racketeer, the chiseling service station owner, and the conniving driver. The entire OPA enforcement staff added up to less than one man for each county in the United States, and since there were a number of other rationing rackets, not all the available inspectors could concentrate on the black market in gas. OPA Administrator Chester Bowles was finally obliged to issue a national appeal reminding citizens that the "lives of our boys in uniform depend on millions of gallons of gasoline."

Indeed, things got so far out of hand that a thriving trade also developed in used ration stamps. Once turned in by stations and garages, they were supposed to be burned. But very few of them, as Valachi recalls, "went into the fire." The Cosa Nostra simply infiltrated the incineration centers with men who retrieved the stamps so they could be put back into circulation.

Of Cosa Nostra members in the racket, only the most affluent could afford the cash outlay—as much as $250,000—for blocks of stolen stamps. These stamps were then bought by middlemen like Valachi and Luciano, who sold them primarily to gas stations and garages. These retail outlets completed the circle by handing in the stamps to the OPA to cover their own black-market sales of gasoline to the public.

The middlemen did not necessarily have to belong to the Cosa Nostra. The great advantage of being a member, however, was that one with a good credit rating did not immediately have to plunk down cash for the stamps. He could, as Valachi did, pay for each allotment out of his profits. He learned later that this was the reason why Luciano had asked him to become a partner. "They—meaning the boys—did not think too highly of Frank," Valachi says, "as far as him meeting his debts is concerned, but everybody knows Joe Cago will always pay what he owes."

A number of stamp heists took place in New Jersey and Valachi usually got his quota from Settemo (Big Sam) Accardi, a powerful figure in the Newark branch of the Cosa Nostra, who is now in prison on a narcotics conviction. Another source was Carlo Gambino, then a lieutenant in one of the New York Families and currently its boss. "Big dealers" like Accardi and Gambino, according to Valachi, controlled the market. As he notes, they did not sell to "everybody." The idea, of course, was to maintain enough tension between supply and demand to keep the stamps moving steadily at the best possible price. "Now maybe they have too many stamps," Valachi told me, "and it will break the market if they let them all go. So they will lie to some guy who is trying to make a buy. I'll give you an example. They will say to him that they ain't got no more stamps and to see Joe Cago. Then I will get a call advising me how much I should charge the guy for the stamps and how many I can sell him. That way they hold the price and we all benefit."

Absolute control of the stamp flow, however, was not always feasible. The gas stamps issued by the OPA bore serial numbers which were good only for specific time periods. Occasionally a robbery would result in a quantity of stamps with a relatively short negotiable life. In such instances, the Cosa Nostra would double-cross the rest of the underworld without hesitation. "It was easy," Valachi says. "They would dump the stamps on the in-between guys who ain't members. They called all these guys, and they would tell each and every one of them, 'Listen, we just have a small amount, but we will let you have them.' Now you must understand that each guy thinks he is the only one, and naturally he breaks a leg to get the stamps."

One memorable morning ration stamps representing 8,000,000 gallons of gasoline were unloaded in this fashion in the New York area within an hour. By nightfall the market price per stamp had plummeted from nine cents to two cents. "Well," Valachi recalls, "a lot of guys took a bath. But a deal is a deal, and they are stuck. Anyway, what could they do against this Cosa Nostra?"

As a rule Valachi could expect to make a profit of one-third on his sale of stamps. A typical transaction with Accardi would always begin the same way:

I will call Sam—Jesus, my phone bill at the time was up over $100 a month—and I will say, "Do you have something on hand for me?" and if he has something, he will tell me, "I got some B-2's."

"What's the market?"

"The market is eight."

"How much will you charge me?"

"For you, Joe, six."

So my profit will be two cents a gallon, and Sam will give me 500,000 gallons. Naturally I try to have Sam take off another eighth [of a cent] at his end, but really it is just play, as I am happy with the price. Sam will say, "The market will hold. Be satisfied."

"Okay," I say, "I will pick up the package this afternoon."

Now we have the stamps on credit, and we owe the guy in Newark, meaning Big Sam, $30,000. But I hustle around, and my partner Frank Luciano has his customers, too. We peddle the stamps in a few days and have a nice profit of $10,000 after paying back Sam. That was all there was to it. If you want to know how we got rid of the stamps, it was mostly to garages. Boy, they were glad to get them. They needed the stamps to cover all the gas they were selling without them. It was the best business I was ever in, some of the big dealers made millions out of it, and it lasted right up to when they threw that A-bomb on the Japs.

Except for a brief period, thoroughbred racing continued throughout the war, and Valachi, in excellent financial shape because of his stamps sales, was able to return to the track, his most treasured acquisition now a light-brown gelding named Son of Tarra. He also used some of his profits to purchase another restaurant, the Aida, on 111th Street and Second Avenue, not far from where he had been born—"Believe me, it was a fine eating restaurant; the head chef cost me $250 a week, and I even had a second chef at $175"—when he received a telephone call one night early in 1945 that would have profound implications for the Cosa Nostra in general and Valachi in particular.

"Hey, Joe did you hear the news?"

"No."

"Vito's coming back!"

9

The circumstances surrounding Vito Genovese's return to this country were as bizarre as the reason for it—the 1934 murder of Ferdinand Boccia. A companion piece to the Boccia slaying, you will remember, was the inept attempt by Ernest (The Hawk) Rupolo to do away with Boccia's friend, William Gallo. When last seen on these pages, Rupolo had been sent to prison for his part in the affair. He was paroled in 1944, but Rupolo apparently carried around his own built-in banana peel.

Within weeks he was involved in another shooting, and once more his intended victim lived to identify him. Understandably dismayed at the immediate prospect of being put behind bars again, Rupolo now decided to reveal the role Genovese played in Boccia's death in the hope that this might help him avoid another prison sentence. Since in the absence of any physical evidence New York State requires a corroborating witness who had nothing to do with either planning or carrying out a specific crime, Rupolo's testimony by itself was not enough. But as he dickered for his freedom, Rupolo came up with such a witness who was present when Boccia was killed. The witness was a small-time hoodlum named Peter LaTempa, by odd coincidence the same man who had knifed Valachi so many years before in Sing Sing.

LaTempa under pressure reluctantly confirmed Rupolo's story, and Genovese was indicted for the Boccia murder. At the time it seemed academic since nobody had the faintest idea where the long-absent Genovese was, a fact that had greatly encouraged Rupolo and LaTempa to talk in the first place. What might have been—how many lives, especially Valachi's, might have ended up differently—had it not been for a former Pennsylvania State College campus cop, then in the U.S. Army's Criminal Investigation Division in Italy, is beyond measure.

The CID man, Sergeant Orange C. Dickey, had been assigned to look into black-market operations behind Allied lines in southern Italy. In early 1944, during a routine interrogation, an Italian suspect boasted that he had nothing to fear since he was under the protection of "Don Vito Genovese," whom he described as someone with powerful connections in the Allied Military Government. The name meant nothing to Dickey, but it stuck in his mind because of the obvious respect in the suspect's voice when he uttered it. Then, in May, Dickey broke up a gang of Italian civilians and Canadian deserters who were helping themselves to huge quantities of U.S. Army supplies and channeling them into the black market. One of the Canadians identified Genovese as the ring's mastermind. In all the confusion of the war it was extremely difficult to trace people, but finally on August 27, 1944, acting on an informer's tip, Dickey caught up with "Don Vito" and had him jailed.

Dickey next paid an eye-popping visit to Genovese's apartment in Naples. As he would later report, he had never seen such a lavishly furnished residence in his life or so extensive a wardrobe for one man. Even more intriguing was a pile of highly priced permits allowing Genovese unlimited travel in occupied areas of Italy, as well as letters from a number of American officers extolling his virtues. One typically declared that Genovese was "absolutely honest and, as a matter of fact, had exposed several cases of bribery and black market operations among so-called trusted civilian personnel." This was in particular reference to his service as an interpreter to the Allied military courts when Naples was first taken, an opportunity Genovese had actually used to eliminate any competition to himself in the black market.

At this point Dickey still had no idea who his prisoner really was. But Don Vito had apparently made some enemies during his tenure as a Mussolini favorite. Two days after his arrest, another informant slipped Dickey a treatise on American crime, written in the 1930s, which featured a photograph of Genovese and identified him as a "notorious gunman." When Dickey showed him the photograph, Genovese—unaware that barely two weeks before he had been indicted back in Brooklyn for the Boccia killing—readily grunted, "Yeah, that's me. So what?"

A less conscientious person would have let it go at that, but Dickey forwarded notification of Genovese's apprehension to the FBI in Washington in case there were any outstanding charges against him. It was not until the end of November that he received word that Don Vito was wanted for murder. In the meantime, enormous pressure had been applied on Genovese's behalf in Italy. Not only had nothing been done about bringing him to trial for his black-market activities, but Dickey had been ordered by his superior officers to remove Genovese from a military prison and have him either post bond or placed in a civilian jail. Determined to hold him in custody until he got some word from Washington, Dickey transferred Genovese to the most secure Italian facilities he could find and kept a close watch on him.

Then, in December, word came from Brooklyn that a warrant for Genovese's return would be forthcoming. It never arrived, however, and nobody in the Allied Military Government, it seemed, wanted to get involved. Dickey might as well have had a leper's bell around his neck. He went so far as to journey to Rome to see Lieutenant Colonel Charles Poletti, chief of the U.S. sector of the AMG, but Poletti, who had been Lieutenant Governor of New York under Thomas E. Dewey before going into the Army, refused to discuss the matter. Dickey even tried Brigadier General William O'Dwyer, in Italy on leave from his job as Brooklyn District Attorney, to no avail. O'Dwyer told Dickey that the Genovese case was of no "concern" to him. Back in Naples, Sergeant Dickey was informed by his superiors that they, in effect, were washing their hands of the whole business. "I was told," as he put it, "that I was on my own, to do anything I cared to."

He decided to bring Genovese back to the United States to face trial. When Genovese learned this, he offered Dickey, whose Army pay was $210 a month, $250,000 in cash if he would just "forget" about him. "Now, look, you are young," Genovese told the twenty-four-year-old investigator, "and there are things you don't understand. This is the way it works. Take the money. You are set for the rest of your life. Nobody cares what you do. Why should you?"

During the winter of 1945, after it had dawned on Genovese that Dickey was not going to cooperate, his manner turned ugly. He told Dickey that he would live to "regret" what he was doing. Then, just as Dickey had completed arrangements to escort his prisoner back on a troop ship, Genovese's demeanor abruptly changed. "Kid," he said, "you are doing me the biggest favor anyone has ever done to me. You are taking me home. You are taking me back to the U.S.A."

The reason for Genovese's surprising switch was quite simple. Having failed to prevent his return from Italy, the Cosa Nostra now moved swiftly. The key witness against Don Vito, Peter LaTempa, had requested protective custody as soon as he found out that Genovese had been arrested. It did him little good. Held in a Brooklyn jail, LaTempa was in the habit of taking pain-killing tablets to quell a stomach disorder. On the evening of January 15, 1945, LaTempa swallowed some tablets in his cell and went to bed. He never woke up. An autopsy showed that he had enough poison in his system "to kill eight horses."

How the poison got into LaTempa's bottle of tablets was never established, but it effectively ended the proceedings against Genovese. After Dickey had deposited him in Brooklyn, Assistant District Attorney Julius Helfand tried for nearly a year to dig up more evidence. Finally he had to go into court on June 10, 1946, to announce that he did not have a case. "Other witnesses [besides LaTempa] who could have supplied the necessary corroboration," he noted, "were likewise not available because they were missing or refused to talk and tell what they knew of this crime because of their fear of Genovese and the other bosses of the underworld, knowing full well that to talk would mean their death."

The presiding judge, in ruling that Rupolo's testimony by itself was not sufficient, told Genovese, "I cannot speak for the jury, but I believe that if there were even a shred of corroborating evidence, you would have been condemned to the chair." Those present in the courtroom observed that the defendant, who seemed annoyed at the formalities holding up his release, smiled slightly during the judge's talk and greeted prosecutor Helfand's remarks with a "stifled yawn."

(As a reward for his information, however useless now, Rupolo was given his freedom, although he was warned that he was virtually committing suicide in electing to leave prison. "I'll take my chances," The Hawk said. For years he led a terror-filled existence, never knowing when his moment would come. Then his tightly bound corpse broke loose from some concrete weights and bobbed to the surface of New York's Jamaica Bay on the morning of August 27, 1964. He had been missing for approximately three weeks. His mutilated body contained a number of ice-pick wounds in the chest and abdomen, and the back of his head had been blown off. In 1967 four members of the Cosa Nostra went on trial for killing Rupolo because of his "activities" as an informer. All four were acquitted.)

Valachi did not see Genovese at once after the dismissal of the Boccia indictment. "I had," he recalls, "some trouble of my own, and I didn't want to go running right away to Vito with it. He is just back from Italy and has beaten this murder rap, and I don't want him to think I am taking advantage of him. Anyway, who knows what he was going to be like after being away all those years?"

Valachi's "trouble" was a problem for the Cosa Nostra judicial process. He had used his fists against another member and, as a result, faced his most serious table to date. If this seems anomalous in the violent world he lived in, Valachi has a ready explanation for it:

It is a hard rule in this thing of ours from the days of Mr. Maranzano that one member cannot use his hands on another member. In

New York the no-hands rule is most important. It ain't all peaches and cream like in Buffalo, say, or them other cities where there is only one Family and everybody is together. It is different in New York. In New York there are five Families—really you must say there are six because when you mention New York, you got to mention Newark, New Jersey—and in New York we step all over each other. What I mean is there is a lot of animosity among the soldiers in these Families, and one guy is always trying to take away another guy's numbers runner or move into a bookmaking operation or grab a shylocking customer. So you can see why it is that they are strict about the no-hands rule.

I was always careful to observe it even when I broke up with that dog Bobby Doyle around 1940, after I caught him sneaking up to my wife and sister-in-law and telling them how I was hanging around with girls every night. Also there were a lot of members who tried to act tough, but in my book they are yellow. There was this time I am in a bar drinking with a girl when Joe Stutz, real name Tortorici, walks in. He is with Trigger Mike Coppola's crew, and first he tells the girl to beat it, and then he says to me that Mike is outside in a car and wants to see me. I find out later Mike was outside, but Joe Stutz don't have to go about it like that. It was a cheap move. He is trying to make me look bad with the girl, so I say, "If Mike wants to see me, he knows where I am. Good-bye."

That is an example of how some people will try to push you. I felt like rapping him in the mouth, but I will wind up on the carpet. Right? I took care of Joe Stutz my own way. He has this guy Patsy running a pizza place for him. Patsy is just a guy, not a member, but he has ideas. I go into the joint with this same girl and I say, "Any room?"

He says, "No," but a little too smart.

So I say, "Come here." Now this is the kind of place where you got to go down a couple of steps to get to the tables. Patsy is just at the bottom step, starting to come up, when I kicked him right in the belly. He is already puking before he hits the floor, and he is out cold. I say to the girl, "Let's take a walk. The food here is lousy."

The next day Joe Stutz is on the phone. "What's the matter with you?" he says. "You know I take care of the guy."

"Well," I said, "teach him some manners. Teach him how to respect people. He shouldn't think that just because you're behind him, he can abuse the whole world. He's been picking up some bad habits from somebody. Do me a favor. Tell him you ain't the only tough guy around."

The incident that finally caused Valachi to break the no-hands dictum would have, by his account, provoked a saint. In 1945, as he had with the Paradise, he sold the Aida because the neighborhood was changing. Soon afterward, Frank Luciano, his partner in the gas stamp racket, invited him to join forces in a new restaurant called the Lido in the Castle Hill section of the Bronx. The liquor license was in the name of Luciano's son, Anthony, since he had no police record at the time. Valachi invested $15,000 as his share, and 250 customers turned out for the grand opening in the winter of 1946. With the approach of spring the average weekly take was around $2,500, and Valachi was delighted. "It was," he says, "almost too good to believe."

The first small cloud on the horizon appeared when Luciano kept putting off Valachi's request that they start drawing salaries on the grounds that they had to build up substantial capital reserve. "Well," Valachi told me, "I was getting worried. Here it is a month into the baseball season, and I ain't seen a dime." He began to investigate, but gingerly, he notes, afraid to hurt Luciano's feelings if his suspicions proved ill-founded. Then one afternoon he bumped into a Bronx bookmaker who remarked, "Hey Joe, I like that partner of yours and his son. They lose with style."

"They been taking much of a bath?"

"Yeah, first the horses, and now every day it's baseball."

That night, in a chat with his partner, Valachi pointedly brought up the coincidence of Luciano's gambling losses and the fact that he had yet to pocket any money from the restaurant. Luciano reacted indignantly, mumbling that he had a good "rabbi," Cosa Nostra lingo for a lawyer at an underworld table, if Valachi wanted to pursue the matter.

"Listen, Frank," Valachi shot back, "we've been in business a long time what with the stamps and everything and we ain't ever had trouble, I'll say that. But if I find something funny, you are going to need more than a rabbi. I'm advising you, take heed."

The warning apparently had little effect. The next evening Valachi arrived at the Lido just in time to catch Luciano helping himself to a roll of bills from the office safe. Valachi grabbed him and said, "Frank, you got to be kidding."

Luciano frantically tried to pull away. "I got to have it," he shouted. "I dropped a bundle on the Yankees. Don't worry, I'll get it back."

Even as Valachi threw his first punch, the thought flashed through his mind that Luciano was purposely setting himself up for a beating. But he was so enraged now that he could not stop himself as he pumped lefts and rights into Luciano's face. Finally Luciano pulled free and ran into the cellar, Valachi right behind him. He cornered Luciano again and continued to pound him methodically until he slumped to the floor. "I got his blood all over me," he recalls, "and I swear I would have killed him if the pieman—meaning the guy who made the pizzas—don't grab me."

Valachi splashed a bucket of water on his unconscious partner and went back upstairs to clean himself up. A few minutes later Luciano appeared, one eye already swollen shut, his nose smashed in, and staggered past him without a word. When he had reached the door, he turned toward Valachi and snarled, "Wait here, you cocksucker. I'll be back."

To Valachi this simply meant that Luciano would be returning with a gun. Like most Cosa Nostra members, Valachi never carried a pistol unless there was a specific reason for it, so he quickly called a "couple of guys" to come to the Lido with one for him. And as he sat waiting for Luciano, he had a moment to reflect on the irony of his situation. "From being right about this whole mess with Frank," he remarks, "all of a sudden I'm wrong. I'm in the right because he has been robbing me blind. Now I'm in the wrong by hitting him. I had violated the rule, and if there is a table, it is going to be for me."

Valachi fully expected Luciano to try to kill him since, by Cosa Nostra standards, he had sufficient provocation for it. But to his astonishment, after a tense hour or so, Luciano telephoned him and said, "Look, I'm sorry. I am willing to forget everything. Let's say it never happened."

Ostensibly agreeing to this, Valachi nonetheless suspected treachery. "I think to myself," he recalls, "that really the best thing is to forgive and forget, but maybe Frank has something else in mind. If I don't say nothing about this, and he does, I'm in even more trouble. So I decide I ain't going to take no chances. I'm going to outsmart Frank if he is trying to trick me."

Thus Valachi went by the book and contacted his lieutenant, Tony Bender. "Tony," he said, "I must see you. I just worked over Frank Luciano." Bender set up a meeting the same night in Greenwich Village at the Savannah Club, which, according to Valachi, he owned jointly with Vito Genovese. There Valachi related what had occurred, including Luciano's offer to disregard the beating. "Okay," Bender told him, "we will play it that way. I won't say anything, but if anyone calls me about Frank, I will say, 'Gee, I meant to talk to you about this. I know all about it from Joe.'"

A few days later Valachi learned that he had correctly gauged Luciano's duplicity. Bender telephoned him and said, "Well, you are on the carpet. There is going to be a table. Frank has reported that you beat him. It will be at Duke's. I'll let you know the date." Valachi was less than comfortable having to depend on Bender for support at the trial. But a rising young hoodlum named Vincent Mauro, whom Valachi had proposed for membership in the Cosa Nostra, was able to relieve him on this score. Mauro had subsequently become a Bender favorite, and he told Valachi that the wily lieutenant, this time at any rate, was being "sincere." "You are lucky," Mauro said, "as nobody likes Frank Luciano. What did you get mixed up with him for?"

"I wish I knew," Valachi said. But he remembers thinking at the time, "It showed how this thing of ours was going to the dogs. Who would think one member would steal from another member like he done?"

Even though he had admittedly "goofed," as he put it, Valachi now felt fairly hopeful that Luciano's thievery would take much of the sting out of the beating charges. Then a new and dangerous element entered the picture. The table had been postponed several times because of the illness of Luciano's lieutenant, Joseph (Staten Island Joe) Riccobono. Luciano belonged to the Vincent Mangano Family, whose underboss was the dreaded Albert Anastasia, and word was relayed that Anastasia himself would appear in Riccobono's place.

The news had, to put it mildly, an unsettling effect on Valachi. In the savage world of the Cosa Nostra, Anastasia had a reputation for incredible ferocity. He lived in a walled mansion in Fort Lee, New Jersey, with a spectacular view of the Hudson River; he was, among his various rackets, absolute ruler of the Brooklyn waterfront and is credited by police with either personally committing or directing literally scores of murders—by gun, by knife, and by strangulation. Worse yet, from Valachi's standpoint, Anastasia was as unpredictable as he was bloodthirsty. "Now I got to worry," he says, "and who can blame me? Everybody knows that Albert is a mad hatter. With him it was always kill, kill, kill. If somebody came up and told Albert something bad about somebody else, he would say, 'Hit him, hit him!' At the table there was no way of telling how he would be."

(Valachi cited a number of examples of Anastasia's handiwork. One of the most horrifying, all the more appalling because Anastasia gained nothing by it, was a celebrated killing that had baffled the police for years until Valachi talked. In 1952 a young Brooklyn resident named Arnold Schuster was en route home when he suddenly spotted a familiar face that he had seen on a "wanted" circular. It was Willie Sutton, the legendary bank robber and the object of a nationwide manhunt. Schuster told the police, and Sutton was arrested. As a result, Schuster experienced a brief moment in the limelight—and then was shot down in the street on March 8, 1952. It was especially mysterious since Sutton was known to be a loner without affiliation in the organized underworld. According to Valachi, however, Anastasia happened to be watching a television news show when Schuster was being interviewed. Suddenly Anastasia exploded. "I can't stand squealers," he told one of his gunmen. "Hit that guy!" To cover himself, Anastasia then had Schuster's killer Frederick Tenuto, murdered in turn. At the time Tenuto was being sought by the FBI for breaking out of prison. His body has never been found.)

When Valachi arrived at Duke's restaurant, his spirits lifted on being told that Vito Genovese was in a room upstairs, and for a moment he forgot about Anastasia. Tony Bender, in bringing him to the table, reminded him soon enough. "Remember," Bender whispered, "for Christ's sake, don't say anything while Albert is talking. You know how he is, so hold your tongue."

Luciano, Anastasia, and an Anastasia lieutenant, Anthony (Charley Brush) Zangarra, were already at the table when Bender and Valachi sat down. Anastasia assumed complete command and, as Valachi recalls, started "lacing" into him at once. "What the fuck's the matter with you?" he snapped. "After all, you been in this life of ours for twenty years. There is no excuse for you."

With his worse fears seemingly realized, Valachi tried to interject, "Albert, I—"

"Shut up. Like I said, you should know better. A rule is a rule. You know you can't take the law in your own hands. You know you could start a war with the kind of thing you pulled."

"But, Albert, this guy was clipping me bad. He put the place behind about $18,000."

"That's what I'm trying to tell you," Anastasia said. "From right you wind up wrong."

Luciano chose this moment to defend himself, and Anastasia abruptly turned on him. "What I want to know from you is what kind of shape the joint is in?"

"It's in bad shape," Luciano replied.

"Why is it in bad shape?" When Luciano hesitated, Anastasia said, "Okay, I already looked into this, and I know what's been going on. You're lucky Joe took a swing at you. Now I have had enough of this. Let's make it short. You two guys are to split up. You can't be together no more. I rule that Joe gets the joint. Frank, how much do you have in it?"

"Fifteen thousand."

"Albert," Valachi said, "I don't want to pay him no $15,000 considering all the money he took out of the business."

"I understand that," Anastasia said. "Nobody said you had to pay him that. Give me $3,500, and the joint is all yours." When Luciano started to protest, Anastasia cut him short. "Frank, I have decided. Take what I allow, or take nothing."

Emboldened by this, Valachi sought to round out his victory. "Albert," he said, "what about the license? The license is in the kid's name; you know, Frank's son. Without the license I got nothing."

"Yeah, I forgot about that. It's good you reminded me. From now on, Frank, you see that your son keeps that license up there. It stays there until that place goes down to the ground. As long as Joe wants that license, he has it. Remember what I'm saying. I hold you responsible if anything goes wrong."

After Anastasia left the table, a chastened Luciano said to Valachi, "When do I get my money?"

"If there's anything left when I finish with the bills, you get it right away. Otherwise, you wait. Now stop bothering me. Every time you open your mouth I feel like rapping you again."

Then Valachi went upstairs to talk to Vito Genovese for the first time in a decade. They greeted each other with studied casualness. "Hey, boss," Valachi said as they shook hands. "It's good to see you. You look good."

"I feel good. How's Mildred and the kid?"

"They're fine, couldn't be better."

"And you?"

"Well, I just got through a table."

"I know. How did things go?"

"Fine. Everything has been worked out."

"Are you short?"

"Well, the place is in bad shape. This dog stole around $18,000 to $20,000. I don't know. I may need some cash."

Genovese turned to one of the men in the room with him, Salvatore (Sally Moore) Moretti, and said, "You heard him. Lend Joe whatever he has to have."

This uncharacteristic offer of financial support was part of an overall effort begun by Genovese to regain the loyalty of the soldiers in his old Family. Still, he did not act immediately to displace Frank Costello as acting boss. There loomed over everyone's head, for instance, the uncertain shadow of Charley Lucky Luciano,* who had never officially relinquished his post. And during this period Luciano, having been deported to Italy, suddenly popped up in Havana, his Italian passport and Cuban residence visa in perfect order. Luciano in Italy was one thing. Luciano in nearby Cuba, however, was something else again, and a parade of Cosa Nostra chieftains shuttled back and forth from the U.S. mainland to huddle with him. How far Luciano would have gotten in dominating the Cosa Nostra again from his new base will never be known, but he put up quite a fight to stay there. He spread so much money in the right places that not until Washington threatened to cut off all legitimate shipments of medical drugs to Cuba did Havana reluctantly agree to pack him back to Italy.

---

*No relation to Valachi's Lido partner, Frank Luciano.

Even then, after Charley Lucky had signaled his support of Genovese, Don Vito had a problem as a result of his long absence. Costello, while he had not been a forceful administrator of Family affairs, was held in high regard by most of his lieutenants, as well as other Cosa Nostra bosses, because of his shrewd moneymaking talents, and while he may have been a remote figure to his soldiers, he solidified his power not only by acting as a financial counselor to the underworld aristocracy, but cutting key members of it into some of his ventures.

Thus in his own Family, Genovese at first could only count on two of its six crews for support in a showdown, those headed by Tony Bender and Michele (Mike) Miranda. Outwardly he kept his composure. Privately, according to Valachi, he raged at the position he found himself in. Valachi recalls one meeting he attended when Genovese whirled on Bender and shouted, "You let these people sew up everything."

"You told me to lay low," Bender replied.

"I didn't tell you to let them bury you!"

"Well," as Valachi observed, "it is easy to see that sooner or later there was going to be plenty of trouble. Vito is building up to something. We just got to wait for him to make his move."

Save for this, however, it was a relatively stable period for Valachi. The Lido was doing well, his horses were winning, he still had his dress factory, his shylocking brought in a steady income, and having achieved a reconciliation of sorts with Tony Bender after his table, he became a partner with Bender and Vincent Mauro in a profitable jukebox operation called Midtown Vending Inc. "It was Tony's idea that I go in with them," he explained, "because he knows I am an old hand with the machines and could hustle a route—meaning locations—pretty good. Well, I accepted as, after all, time heals everything."

The 1940s closed with a notable change in Valachi's mode of living. Since his marriage he had resided in various apartments in the Bronx. Now Valachi joined the ever-fashionable flow of mobsters to the suburbs. Like his colleagues, he never engaged in anything remotely connected with the underworld in his new surroundings. Indeed, when Genovese heard of his prospective move, he drew Valachi aside for some words of advice. "It's different from living in the city," Genovese said. "Make the people in the neighborhood like you. Don't fool around with the 'weak' [ordinary, law-abiding citizens]. Give to the Boy Scouts and all the charities. Try to make it to church. Don't fool around with the local girls."

His wife, Mildred, was responsible for the move:

Around 1950 or so Mildred decided that nothing will do except that we had to own a house. One day she calls me and she says she has found a house in Yonkers, New York. She wanted me to see it. I told her if she liked it, that was good enough for me. She said the house would cost $28,000, so I gave her $5,000 to make a deposit.

By this time the boy had finished his days in school. It was one of the best in New York City. Its name was Mount St. Michael. As I remember, it cost about $1,600 or $1,800 a year to keep him in that school. He boarded there and came home only on holidays, as we wanted to keep him off the streets of the Bronx.

When he had finished, I asked him if he wanted to go to any other school, and he said no, he wanted to go to work. Well, the kid took up mechanics, but he didn't do too well at that. So I got him a good job. I'm not going to mention it, but he can have it for life, and the last I heard he is doing all right. He married young, and I built three rooms onto the house in Yonkers for him and his wife.

The three rooms that I added cost about $10,000. I spent another $2,500 for awnings, and I had to do a lot of work on the grounds. A houseowner will know what I mean. I spent quite a few dollars putting in a fence, trees, and whatnot, and also a concrete driveway so as not to mess up the house from the dirt road. I painted the whole house inside and out, and I used the best paint. I was very handy at home, that is, when I was home, as I was always busy with this and that. All in all, I'd say the house cost me over $40,000. Well, it was worth it. It was a beautiful corner house. My address was 45 Shawnee Avenue. As far as the neighbors are concerned, I was always a gentleman. Naturally by the hours I kept they got around to asking Mildred what did I do for a living. Mildred told them that I had the Lido, so they figured I was just a guy who ran a restaurant. Some of the neighbors would come into it every now and then, and they all said they liked the food very much. I ran a clean place. Any girl could come in there alone—you know, without being escorted—and if some guy bothers her, he gets thrown right out on his ass.

I'll say one thing thinking about them days. I am a happy man that I brought up my kid naïve, so he wouldn't be in the life I was.

Despite this surface tranquillity, Valachi's career would soon enter its most turbulent phase since the Castellammare War tore the Cosa Nostra apart. Not only did "trouble" erupt when Genovese at last made his bid for boss, but his mere presence on the scene seemed to trigger all sorts of explosive side effects:

It was a bad time for us. Everyone was a little nervous. I felt that at any moment I could get hit with a shotgun blast. I would come home in the wee hours, and it would be very quiet, and I would stand in front of my door sometimes, and I would hold my shoulders tight expecting a blast. I did not carry a key for the house, as I used to lose them. It seemed like hours, those few minutes I would have to wait until Mildred opened the door. But I must say other times I just didn't care, and I would stand there relaxed. What could I do about it anyway?

10

One of Frank Costello's most powerful supporters in the old Luciano Family was William (Willie Moore) Moretti. Costello and Moretti had been born barely a block apart in East Harlem and had remained close through the years. In the 1930s Moretti, as a lieutenant, had moved out of New York proper and headquartered himself in northern New Jersey. There he supervised the operation of a variety of rackets that had grown out of his initial success as a bootlegger and later as a numbers kingpin. "Willie had lots of men," Valachi says, "about fifty or sixty men in Jersey, throughout Jersey. He had some in New York too, but Jersey was his stronghold. Most of them were members, and some of them were not—he had a lot of things going for him—but they were all with Willie Moretti. He was like independent. He had his own little army. That was the way we expressed it among ourselves. That was the way we thought."

Moretti, however, had a personal problem that was becoming a matter of concern to his fellow chieftains in the Cosa Nostra. He was suffering from advanced syphilis, as had Al Capone, and it was getting to his brain. It had not completely overwhelmed him yet; there were times when he was rational as ever. But there were other intervals when he babbled on crazily about things better left unsaid. During these lapses Costello sequestered him under medical care. With Genovese circling around him, he needed Moretti more than ever for his symbolic support, if nothing else. This shepherding of Moretti reached its heights during the national crime hearings conducted by Senator Estes Kefauver. Moretti, who had been subpoenaed, was in a bad way, and Costello kept him on the move along the Cosa Nostra network all through the Far West, accompanied by a doctor and a male nurse. Only when Costello was certain that he had come out of it did he permit Moretti to return East to answer questions. Even so, while he did not say anything especially damaging, he often seemed on the verge of it in the course of his ramblings on the stand. The situation grew more serious in a subsequent appearance before a New Jersey grand jury. Worse yet, he grew increasingly garrulous with reporters, apparently enjoying the attention and the opportunity to reminisce about the "old days."

Now, according to Valachi, Genovese used this in a cunning ploy to undercut Costello:

Vito is like a fox. He takes his time. He wants everything to be legal. He starts talking it up, first among us that are closest to him, which is Tony Bender's crew, so then we will talk it around, and the word will spread, and other members will start thinking about it. What Vito says is that Frank Costello is right about a lot of things, but he is wrong about this. He says it is sad about Willie and that it ain't his fault. He is just sick in the head, but if he is allowed to keep talking, he is going to get us all in a jam.

You see how Vito is agitating to get Willie Moretti killed, and it ain't making Frank Costello look any too good. One time Vito has this meeting with some of us downtown in the Village, and he says, "What are we, men or mice?" Well, he is scheming like this, saying Willie has got to be hit because he is not well. I remember him telling us, "He has lost his mind and that is the way life is. If tomorrow I go wrong, I would want to be hit so as not to bring harm to this thing of ours." Then, after the seed has been planted, it naturally grows, and there is agreement in the *Commissione*—meaning all the bosses—that Vito is right.

The contract to murder Moretti was "open," with no specific member of the Cosa Nostra assigned the task. "They will take their time about it," Valachi explains. "They are not going to worry about this week or next. Whoever has the opportunity, more or less, will hit Willie."

The opportunity, as it happened, came to Valachi's former shylocking partner, John (Johnny Roberts) Robilotto, who was now with Albert Anastasia. Robilotto and a "couple of other members," according to Valachi, had a morning appointment with Moretti in New Jersey. "It wasn't at Duke's," he says. "It was about seven doors away on the same block. I forget the name of the joint. Anyway, that's when Willie gets his, from Johnny and these other people."

Valachi heard about it over the radio. He had slept until early afternoon, as was his habit, and then left for the Lido. He was looking for a new waitress, and on the way he stopped by a girl's apartment to see if she would be interested. He was about to go, after being told by the girl that she already had a job, when a news broadcast summarized the killing. Valachi promptly stayed put in the apartment. "Even if it was all arranged and everything about Willie," he says, "you must be careful. Willie was pretty well liked, and that's why Vito took so much time selling the idea to the old-timers. So you never know, there could be trouble."

He attempted to get in touch with Tony Bender to find out what the situation was. Unable to locate him, Valachi remained away from the Lido until nearly midnight, when he finally contacted his lieutenant. "Is it okay to be seen?" he asked.

"There's nothing to worry about," Bender replied. "Go about your business." Then when Valachi got to the restaurant, he learned that Robilotto had been there waiting for him:

He told the waiter to tell me he was there to celebrate. I understood what that meant. He knew Willie Moretti and I never got along. I'll explain why. Willie helped me out once in the numbers, but he was supposed to do that. I could have made a beef if he don't. But I understood that after Mr. Maranzano was hit and I had to hide out, it was Willie and his pal Ciro Terranova who was behind the idea of dumping me, and I had told Johnny about it.

Now there was a big investigation. It was in the papers about someone in the joint identifying Johnny Roberts and that there were a couple of hats in the joint they were tracing, and the next thing I hear is Johnny has been picked up. Then this detective in the Bronx—at the 204th Street Station—calls me and asks me will I come down, as these two detectives from Jersey want to see me. "Certainly," I say, "why not?"

I go down, and they ask me where I was on such-and-such a day— I don't remember the date, but it was the day Willie was shot—and I tell them. Then they show me pictures of Johnny Roberts and one of the other guys, and they want to know if I know them. I say, "No."

They bust out laughing. "Come on," they say. "We know you know Johnny."

"Listen," I say, "you asked a question. You got an answer. What else do you want to know?"

"We want you to come over to Jersey."

"What is this," I say, "a pinch?"

"No," they say. "We just want to talk some more."

I say, "I ain't going to Jersey. You want me in Jersey, talk to my lawyer." I told them I only come down because of the Bronx guy— meaning the detective—and enough is enough. They said they might want to get in touch with me again, and I said, "You know where to find me," and that's the last I ever seen of them.

A few days after this I saw Johnny. I forget if he was out on bail— it don't matter, as nothing ever came of it—and of course, I ain't going to talk to him in plain English about it. He ain't supposed to tell me nothing, and I got to curb what I say and not put him on the spot. I say, "I am sorry I missed you up at the restaurant that night."

And he says, "Yeah, we would have had a time."

Now there was still a lot of talk around about the hats they found in the joint fitting this one and that one, so I say, "How do you stand with the hats?"

"Don't worry," Johnny says, "it ain't my hat."

Well, Willie Moretti had a hell of a funeral, lots and lots of cars and flowers. Usually when a top guy is deserted—like Mr. Maranzano, or Albert Anastasia, I will get to him later—he is deserted all the way to the cemetery. But Willie was not deserted because it was sort of, as we put it, a mercy killing, as he was sick.

After it was over and done with, Vito Genovese even said, "Lord have mercy on his soul." Naturally I formed my own conclusions. There it is, I said to myself, Vito has found his opening and gone through, and now he's off and running.

(According to New Jersey police records, at approximately 11 A.M. on October 4, 1951, Moretti was found dead, shot twice in the head, in Joe's Restaurant, 793 Palisades Avenue, in Cliffside Park. Three white, male patrons had been seated in the restaurant, one at the counter, two at a nearby table. The man at the counter, tentatively identified as John Robilotto, also known as Johnny Roberts, left the restaurant and returned immediately with Moretti, whom he introduced to the other two men. The only other persons in the premises were the wife of the owner and a waitress. Both women retired to the kitchen. While there, they heard gunfire. Immediately upon coming outside, all the males had fled, with the exception of Moretti, who was lying on the floor. Tentative identification by the waitress was made of Robilotto, although she "could not be sure." Additionally, there were two men's felt hats left at the premises, one bearing a cleaning mark which was traced to a cleaning establishment in New York City on the Avenue of the Americas. This information, however, was prematurely released to the press, and by the time investigators arrived, the slip pertaining to this hat was missing. Robilotto was indicted for murder in June 1952. On October 14, one Joseph Valachi was contacted and questioned, among others, in connection with the slaying by detectives of the New Jersey State Police. Valachi denied ever seeing Robilotto or another suspect in the case, and charges against Robilotto were subsequently dismissed because of insufficient evidence.

In early September 1952, Valachi was summoned by Tony Bender. It resulted in his most important Cosa Nostra contract. They met for dinner, he says, in a Greenwich Village restaurant on Thompson Street called Rocco's. After a few minutes of perfunctory chitchat, Bender got down to business. A soldier in the Thomas Lucchese Family, Eugenio Giannini, had turned out to be an informer for the Bureau of Narcotics. "Gene has been talking to the junk agents," Bender said. "The old man [Genovese] has got the word personally from Charley Lucky. Charley says Gene is the smartest stool pigeon that ever lived. He has been talking to the junk agents for years. He has got to be hit, him and anybody with him."

The news about Giannini was correct. He was a Narcotics Bureau informer. In 1942 he had been picked up on a heroin conspiracy charge and served fifteen months for it. Later, like many other informers used by the Bureau of Narcotics, the FBI, and so on, Giannini moved in a twilight zone, continuing his own underworld activities while passing on choice tidbits about his colleagues from time to time. Furthermore, as was the case with every informant until Valachi, he discussed only specific individuals and crimes, never the Cosa Nostra itself. The idea, from his point of view, was that if he happened to be caught in one of his illegal operations, he could always claim that he was on an intelligence-gathering mission or, if need be, fall back on his previous cooperation—and future potential—to escape punishment. For the law enforcement agency involved, an informer like Giannini always represents a judgment decision: Is the intelligence he supplies worth overlooking what he might be engaged in at a given moment?

In 1950 Giannini had left for Europe with two projects in hand. One was smuggling U.S. medical supplies like sulfa and penicillin into Italy where they were in high demand in an economy still struggling to recover from World War II. Another commodity in equally high demand at the time in both Italy and France was U.S. currency, and Giannini did his best to satisfy it, although his banknotes, of

course, were counterfeit. He intended to use part of the proceeds
from these enterprises to finance the purchase of heroin for distribution, and even bigger profits, back in America. Toward this end, as
he lived it up on the Continent with assorted mistresses, Giannini
made contact with Luciano in Naples and on the side began tipping
off the Narcotics Bureau about various aspects of the deported chieftain's traffic in drugs. Meanwhile, the head of the bureau's European
office, Charles Siragusa, began putting together from other sources a
picture of Giannini's own elaborate smuggling plans, and the problem of what to do with the prize informant was going to have to be
faced fairly soon.

Then there was an unexpected development. The Italian police
suddenly arrested Giannini on charges of dealing in counterfeit
dollars and tossed him in jail to await trial. He managed to sneak
out a letter to Siragusa, graphically describing the filth, heat, flies,
and bedbugs that featured his new surroundings, and demanded
that the Narcotics Bureau arrange his release. At the same time it
was learned that Giannini was still plotting from his cell to ship at
least ten kilograms of heroin into the United States, so the bureau
decided to let him cool his heels for a while. In desperation Giannini wrote another letter to Siragusa, specifically citing the information he had passed on regarding Charley Lucky. While Valachi
does not know how Luciano discovered that Giannini was an
informer, the Cosa Nostra has intelligence sources of its own, and
it is likely that this letter did him in.

Eventually Giannini went on trial on the counterfeiting
charges and, when the chief witness against him suddenly changed
his testimony, was acquitted for lack of evidence. Giannini then
flew back to New York, where the Bureau of Narcotics kept him
under close surveillance.

It was not long after this that Valachi had his meeting with Bender. Ordinarily Giannini would have been a matter for the Lucchese Family to handle, but as part of Genovese's continuing campaign to assert himself, he was, as Bender told Valachi, "anxious to throw the first punch," and the fact that Luciano was the injured party gave him the opening he needed.

Valachi had known Giannini for years. As a matter of fact, Giannini owed him money borrowed just before he had gone to Europe. It was precisely because of this debt, as it turned out, that Valachi was picked for the contract. When Bender told him that Giannini had to be eliminated, Valachi replied, "Well, there goes a couple of thousand that he owes me."

"Yes," Bender said. "I heard it was something like that. You know, Gene is just back from Italy, and he is moving around, and we can't seem to find him. If you hear anything, call me. Don't start thinking about trying to save that money."

The inference was clear, although Valachi went on with the game. "Can't find him?" he echoed.

"That's right. It could be Gene is on to something."

"Well, I'll find him. Does that satisfy you?"

"I will have to talk to the old man about that."

The next night Bender informed him that Genovese was quite pleased with the offer. "Finding" Giannini, in the euphemistic exchange that had taken place, meant killing him. As Valachi explained to me, "I got no choice. I have to volunteer for the contract. If I don't and something goes wrong, they can blame me because he owes me money and they can accuse me of tipping Gene off. That's why I answered Tony the way I did. If they can't find him, I will. Now you know how this Cosa Nostra is."

Valachi's next step was to locate Giannini. As he suspected, it was not very difficult. He simply telephoned him at home at about 10 P.M. Working on the assumption that Giannini would think he was calling about the loan, he spoke just long enough for his quarry to identify his voice and then said, "Meet me on the corner."

The "corner" he referred to was the intersection of Castle Hill and Westchester Avenues in the Bronx near the Lido. Apparently these directions were sufficient for Giannini. He replied, "I'll be over in twenty minutes."

Valachi waited in a doorway in the dark and watched Giannini drive up. Then he saw another car stop down the block. When Giannini started to say something about paying him back as soon as he had some cash, Valachi silenced him and hustled him into a bar on the corner. "Forget about the money," he said. "I think you got a tail. What do you have, agents following you?"

"Jeez, they must be watching you."

"Maybe they are," Valachi said diplomatically. "Let's pass it up for now. I'll give you a ring in a couple of days."

Giannini left the bar the way they had come in. Valachi waited for a moment and then used a side door. Standing in the shadow, he saw his suspicions confirmed; as Giannini drove off, the second car pulled out as well.

The Giannini contract was a classic instance of how the Cosa Nostra power structure removes itself from the actual commission of a crime. The impetus for the murder came from Luciano, who would be, of course, in Italy all the time. Genovese ordered it, but he would be nowhere near the scene when it happened. Nor would Tony Bender, who transmitted the command. Even Valachi, who was responsible for carrying out the contract, would not be physically present. How it was to be done and whom he used was entirely up to him, and he selected three East Harlem "kids," as he called them, rising hoodlums who were in line for membership in the Genovese Family, for the actual execution. Two were brothers, Joseph and Pasquale (Pat) Pagano; the third was Valachi's own nephew, Fiore (Fury) Siano, the son of one of his sisters.

Always painstaking in an affair of this sort, the presence of agents from the Bureau of Narcotics made him infinitely more cautious. A few days after his abortive meeting with Giannini, he called him again and set up a rendezvous at another bar near the Lido, called the Casbah. Valachi brought along one of his three recruits for the job, Joseph Pagano, to, as he says, "kill two birds with one stone." The first reason was to enable Pagano to get a good look at his victim; the second to put Giannini at ease the next time he saw Pagano. There was also a third reason for the meeting. Valachi had purposely not suggested the Lido because he wanted to find out if Giannini was still being followed. Thus he and Pagano waited across the street as Giannini entered the bar, and once more he spotted a tail. Then Valachi and Pagano entered the bar separately. He introduced Pagano to Giannini and said, "Gee, Gene, every time I call you, you got somebody covering you."

Giannini expressed astonishment. "I'm glad you tipped me," he said. "I can't understand it."

"Ah, forget it. Let's have a drink."

The two men chatted on for a few minutes, Valachi asking him about his sojourn in Europe, when Giannini suddenly said, "Jesus, I had the creeps last night."

"What do you mean?" Valachi recalls saying.

"I don't know. It's hard to explain. I feel like I'm going to be killed."

"What are you talking about?" Valachi quickly said. "Why do you say a thing like that?"

"It's just the way I feel."

To break the mood, Valachi called over a girl in the bar who had once worked for him in the Lido, introduced Giannini to her as "my old pal," and bought another round of drinks. Then he drew Giannini aside and said, "Listen, you got to cheer up. Why don't you go out with her and have a good time?"

"Joe, I'm a little short. That's why I ain't paid you. I got a deal working, but I'm broke right now."

Valachi promptly handed Giannini $100. "Go ahead and enjoy yourself."

He stayed for another drink before telling Pagano, "Let's go." Giannini remained with the girl when they left. As they walked to Valachi's car, Pagano said, "I can't get over it, the way he was saying he was going to get hit. Well, it must be true he's talking. Twice you call him, and twice he has a tail."

After making certain that he was not being followed, Valachi dropped Pagano off in East Harlem. The next afternoon he contacted the girl he had left with Giannini and learned from her that the condemned man had mentioned something about a "game" in Harlem. Valachi passed this intelligence on to the Paganos and Siano. A few days later one of them—he thinks it was Joseph Pagano—reported back that Giannini was working "at a drop" on Second Avenue for a dice game around the corner on East 112th Street. "I'll explain what working 'at a drop' means," Valachi notes. "The crapshooters got to this place, which is the drop, before they go to the game. In other words, the game was about half a block away. The fellow who is working the drop looks over the players, and if he sees they are okay, he escorts them to the game at such-and-such address or such-and-such room in a hotel."

Upon receiving this information, Valachi immediately asked, "What about the agents?"

"We didn't see any. He must have beefed about it or something."

"Okay, that's what we have to know."

There was, however, a further complication when Valachi discovered that it was a Cosa Nostra dice game run by a member of the Lucchese Family, Paul (Paulie Ham) Correale. He had previously sent Pagano to Greenwich Village to pick up the guns for the killing from Tony Bender. Now he had to resolve the matter of the site. He found his lieutenant at an after-hours place, the Gold Key Club in midtown Manhattan, which, according to Valachi, was owned by Bender. Since the Gold Key Club was often under police surveillance, Valachi took his usual precautions in going there, parking his car some distance away and switching to a cab to avoid having his license number noted.

Before he could speak, Bender snapped, "What the hell's the holdup? Let's get this thing over with. I hope you're not trying to save your money."

"Tony, listen to me," Valachi protested, "there have been junk agents all over this guy." Then he brought up the problem of the drop. "The game belongs to Paulie Ham," he said. "Is it okay to get Gene there?"

"Well, you're right about this. I'll have to talk to the old man. I will make sure. Call me up here tomorrow night. You don't have to come down. Just call me up, and I will have an answer for you."

Valachi telephoned as directed, and Bender said, "It's okay."

For Giannini now, despite all of Valachi's machinations, time was running out. He would get one small reprieve. Valachi's plan for the assassination was to have one man stationed in a getaway car on 111th Street, a block south of the game. The other two, after completing their mission, were to cut through one tenement on 112th Street, exit out of another building facing 111th Street and into the car. On the night of September 18, 1952, Valachi rode with the Pagano brothers and Siano to inspect the area. Everything seemed set. Giannini was standing on the sidewalk in front of the drop, and no agents could be found in the immediate area. Then, just as he was about to leave them, Valachi recalls asking Pat Pagano if he had made certain that there were no locked doors barring the escape route.

"No, I—"

"That's it," Valachi broke in angrily. "Go home and sleep. Check those halls tomorrow. What's the matter with you kids? Are you crazy? You ain't doing a thing until you're sure. I'm responsible for this."

The next afternoon, assured that this final safety measure had been taken, Valachi gave the go-ahead. He told Siano that he would be waiting for news of what happened at a restaurant on 114th Street and Second Avenue. In telephoning him, he added, Giannini was to be referred to as a girl. Then he reminded Siano that as soon as the shooting was finished, the pistols were to be dropped into the East River off the Third Avenue Bridge.

Valachi, to fix his own alibi, arranged to have dinner with three friends in the restaurant. Around midnight, Siano telephoned and said, "The girl hasn't shown up yet."

"Okay, I'm going up to my place."

Valachi then asked his dinner companions to drive him to the Lido. There, around 4 A.M., as he remembers, he received another call. It was all over. "We saw her," Siano said. "We're going on a trip for a couple of days."

"Fine," Valachi replied and went home to bed.

(New York City police records show that at 6 A.M. on September 20, 1952, the body of Eugenio Giannini, age forty-two, of 282 West 234th Street, was found in the gutter in front of 221 East 107th Street. The cause of death was gunshot wounds in the head by persons unknown. Further investigation indicated that the shooting actually took place on Second Avenue near East 112th Street. The deceased had narcotics arrests both for violation of state and federal laws, and it was learned through confidential sources that he had been an informant for the Federal Bureau of Narcotics.)

At the time the Narcotics Bureau believed that Giannini had been slain not because he was an informant, but because he had tried to bilk his associates out of most of their share of the heroin he was then engaged in smuggling into the country. There was some justification for this. According to the bureau's information, ten kilos were involved in the transaction. Undercover agents discovered, during the course of their investigation, that Giannini had quietly dispatched his brother-in-law to Italy to bring in six kilos on his own. This led to the brother-in-law's arrest in Salerno with the heroin in hand about a month before Giannini was murdered.

Once it had been established that the victim had been taken to 107th Street after being shot, there also was conjecture in the Bureau of Narcotics that this was simply a neat symbolic gesture since, until Valachi pieced together the organization of the Cosa Nostra, what would become known as the Lucchese Family was called the East 107th Street Mob.

Valachi was as curious about the body being removed. The first he heard about it was over the radio when he woke up that morning. The reason for it was not quite so esoteric, and for a while it appeared to an outraged Valachi that he was headed for another table as a result.

"The guys running the game," Genovese told him a few days later, "claim they had to move the body to save it. They are pretty mad. They say it cost ten grand to pay off the cops."

For a moment Valachi thought that Bender had not cleared the site of the killing, but Genovese confirmed that he had given his approval. "I just want to find out what this is all about," he said. "So find out."

Valachi reported back to Genovese full of righteous indignation. "You know those guys that said they took Gene out of the neighborhood to throw the heat off the game? Well, they're lying."

"What do you mean?"

"There were a couple of boys working at the drop with Gene. When they found him, they thought he still had a chance. They were rushing him to the hospital!"

"Is that the way to the hospital?"

"Of course it is," Valachi said. "You don't know the neighborhood like I do. They are going to Fifth Avenue Hospital.* They have to drive down Second Avenue, and naturally they turn at 107th, as it's a westbound street and it takes them right to the hospital. On the way they realize Gene is dead, and they dump him off. Now they want to play heroes. They want everyone to think they knew Gene was dead and they took a chance being caught just to get his body away from the crap game. I won't go for it. Everytime I'm told to do something, its a mess, and I'm in the middle."

---

*Actually Flower and Fifth Avenue Hospital, located at 106th Street and Fifth Avenue.

"Where did you get this from?"

"From the kids, Fiore and the Paganos."

"Well, don't worry, I'll take care of it."

With all of his old mistrust of Genovese flaring up again, Valachi decided to take no chance. He buttonholed Thomas Lucchese, Giannini's Family boss, at "somebody's wake" soon afterward and said, "Tommy, a lot of your boys are acting cold to me about this. I don't care how they feel, but you and me go back a long·way. I want to know how you feel. Do you think I went crazy and did it myself without orders?"

"Forget about it," Lucchese said. "Everything is fine. It wasn't your fault. The guy got what he deserved."

(One of the three men Valachi says he used in the Giannini murder, his nephew, Fiore Siano, suddenly vanished about nine months after it became known that he was talking. According to intelligence gathered by the New York City police, "Siano disappeared about the end of April or the beginning of May 1964. He has not been seen since three unknown males took him out of Patsy's Pizzeria, 2287 First Avenue, during the aforementioned period. Siano is believed dead. The rumor is that his body was disposed of in such a manner as to prevent it from being discovered." Siano liked to shoot pool, and I spoke to one of the players in the last game he was known to have been in. In what is certainly the understatement of the year, the player recalls, "Fiore seemed moody, like he had something on his mind." Of the Pagano brothers, Joseph was sentenced to five years in 1965 for his part in what was described as a classic case of Cosa Nostra infiltration of a legitimate business; Pasquale Pagano, characterized by the Bureau of Narcotics as an "up-and-coming" underworld figure, has been in and out of prison and is currently at liberty. Valachi's testimony against them, without corroboration, is legally insufficient. Indeed, the Giannini murder illustrates the near impossibility of the successful prosecution of a gangland slaying. The police were called immediately after the shooting, but nobody saw anything. An elderly janitor was found cleaning the sidewalk. When a bloodstain still on the pavement was pointed out to him, he said, "Blood?")

Within a year Valachi would be involved in another murder contract with Pat Pagano and Fiore Siano, this one on behalf of Vito Genovese himself. Genovese's climb to power had suffered an embarrassing interruption in December 1952, when his wife, Anna, left their home in Atlantic Highlands, New Jersey, and sued for divorce on the grounds that his cruelty "endangered her health and made her life extremely wretched." Mrs. Genovese, to say the least, was not the typical gangster's wife. In asking for separate maintenance, she wasted few words in court portraying her husband as a savage mob leader with a huge underworld income. Not only did she pinpoint the location of various safe-deposit boxes he kept full of cash both in the United States and Europe, but she detailed, among other items, his vast gambling interests, nightclubs, loan-shark activities, and labor union kickbacks. From one racket alone—the so-called Italian lottery, based on a popular numerical game of chance in Italy—she testified that Genovese took in between $20,000 and $30,000 a week. "I know specifically about the Italian lottery," she said, "because I myself ran the Italian lottery."

As previously noted, the Cosa Nostra to a man was goggle-eyed that Genovese would let her get away with it. But while his love for her apparently stayed his hand, it did not "go so good" for one Steve Franse, who had been a partner first with Genovese in some of his nightclubs and then with Mrs. Genovese in running some others she personally had in the Greenwich Village area which were not, in her words "part of the Syndicate."

Genovese felt that his wife had fallen away from him during his long absence and blamed Franse for not keeping a closer eye on her. Thus early in June over a plate of veal parmigiana Bender informed Valachi that Franse was to be killed. Bender did not use Genovese's domestic problems as the reason for it; instead, he presented the usual catchall charge that Franse was a "rat." While Valachi had known Franse since the 1930s, he had never been to the Lido, and this would be the pretext to take him unawares. "Stay at the restaurant every night after you close," Bender told Valachi, "until you hear from me. If I say go home, go home. If I say wait, wait."

Franse's murder, according to Valachi, was brutal:

Every night, I'd say for sixteen nights, I get a call from Tony and he says, "Go home." Then I get this call, and he says, "Wait!"

So I lock up everything tight and pull the curtains and just sit there. A little after four o'clock there is this knock on the door. I open it, and in comes Pat Pagano and Fiore. They got Steve Franse with them. "Hey, Joe," Pat says, "we want Steve to see your joint."

Well, I fix drinks, and we talk about how the Lido is doing, and we walk around the front, and then we go into the kitchen. That's when it happens. Steve is a little guy, and Pat is pretty big. Pat grabs him from behind—he has got him in an armlock—and the other guy, Fiore, raps him in the mouth and belly. He gives it to him good. It's what we call "buckwheats," meaning spite-work.

I'm standing guard by the kitchen door when Pat lets go and Steve drops to the floor. He is on his back, and he is out. They wrap this chain around his neck. He starts to move once, so Pat puts his foot on his neck to keep him there. It only took a few minutes.

Approximately half an hour afterward, Valachi stepped out of the Lido and checked the street. No one was in sight. He next slipped behind the wheel of Franse's car and started the motor. At this signal, he says, Pagano and Siano emerged with Franse propped up between them as if he were drunk and put him in the back seat. Valachi got out of the car, and the two men drove off in the direction of Manhattan with Franse's corpse.

(New York City police records note that about 9:55 A.M. on June 19, 1953, the body of Steven Franse, male, white, fifty-eight, of 1777 Grand Concourse, Bronx, was found in the rear seat of his automobile parked in front of 164 East 37th Street, New York City. Cause of death: manual strangulation, with contusions and abrasions on face and body, as well as a fractured left rib.)

Out of the swirl of fear and mistrust inside Cosa Nostra ranks brought on by pressure from the Bureau of Narcotics, Valachi now received stunning news about his old pal and mentor, Dominick (The Gap) Petrilli. After serving time for his 1942 narcotics conspiracy conviction, Petrilli, who had neglected to become a U.S. citizen, was deported to Italy, and Valachi had not seen him since then.

Around the middle of November 1953, Tony Bender told Valachi, "The Gap is back. He got picked up in Italy for something and made a deal with the junk agents. They let him come back to set us up."

Petrilli belonged to the Lucchese Family, and Valachi quickly broke in. "I don't care what The Gap's doing. Don't mix me into it. Let his own people handle it. I ain't going to be in the middle again, like with Gene."

"Nobody is asking you to take the contract. He is certain to see you. Just tell me when he does. He will explain that he sneaked in on a boat and jumped off. Be careful he doesn't have a tape recorder when he's talking to you."

Bender's warning left Valachi badly shaken. Recalling his thoughts at the time, he told me, "It seemed like everything got messed up when Vito came back. At least with Frank Costello life was nice and peaceful. First it's one rumor; then it's another. Now The Gap is supposed to be a rat. To tell the truth, I don't want to talk to nobody no more. I'm afraid to hear what the next thing is, and I'll get in trouble just for knowing it."

Late one night, about three weeks after the conversation with Bender, Petrilli came into the Lido. He was drunk or else did an excellent job of acting as if he were. He greeted Valachi with an enormous bellow and threw his arms around him.

Valachi, feigning delighted astonishment, asked Petrilli how he dared to walk around so openly.

"I ain't walking around," Petrilli said. "I just got off the boat, and I got to see you, right? You're my friend, right? I couldn't take Italy no more. I had to get away. I was in the hold of this freighter for twenty-seven days. Twenty-seven days! You don't know what it was like. It cost me $3,000."

Petrilli then told Valachi to arrange a meeting with Vincent Mauro and another soldier in Tony Bender's crew, Pasquale (Paddy Mush) Moccio.

"I'm going to make you guys rich," he whispered. "I got a line on a ton of stuff. Joe, you and me are going to Cuba to pick it up."

Valachi had listened silently to this. Now, to see if Petrilli was in fact carrying a recorder, Valachi suddenly opened his coat, simultaneously exclaiming, "Hey, you have lost weight!" Finding nothing, he lowered his voice and said, "Gap, listen to me. It's dangerous around here for you. If you're going to Cuba, don't wait. Go."

Petrilli gave no indication that he heard him. Instead, he called for another drink. After downing it, he lurched out, saying, "Set that meeting up. I'll be in touch in a couple of days."

Valachi waited an hour and then telephoned Bender. "He was here."

"Is he still there?"

"Why didn't you call while he was there?"

"I told you, I don't want no hand in this."

"Did he say where he was going?"

"No."

"How do you size it up?"

"Well, he said everything you said he would say."

It was the last time he saw Petrilli. On a slow December night Valachi closed the Lido early and went home. He was asleep around 5:30 A.M., when Mildred awoke him. Two detectives were standing beside her. One of them said, "Your wife says you got in at three. How come you're home so early?"

"There weren't any customers. So what?"

"Your friend Petrilli just got hit. What do you know about it?"

(New York City police records show that at about 3:50 A.M. on December 9, 1953, at a bar and grill located at 634 East 183d Street, Bronx, owned by Albert Mauriello, three unknown white men, all armed and wearing dark glasses, entered the premises and shot Dominick Petrilli, male, white, fifty-four, alias The Gap, causing his death. Case active.)

Valachi was questioned once more and denied any knowledge of the murder. But from a source who had been with Petrilli at the time—"He wasn't a member, just a guy around, so why get him in trouble?"—he learned that The Gap had been in a card game at the bar, which was owned by the brother of onetime heavyweight contender Tami Mauriello. Valachi was told that Petrilli was ahead $1,300 when the three killers, members of the Lucchese Family, walked in. He took one look and fled to the men's room, hoping to escape through a window, but was cornered there and had "his brains blown out." The source added a footnote. The players in the game had scattered immediately; one of them, however, managed to run back to rifle Petrilli's pockets before the police arrived.

"Well," Valachi says, "even if I was sure The Gap was a stool pigeon, I wouldn't have done nothing to him. How could I forget how he took me to Brooklyn and kept me out of the way when Mr. Maranzano got his? Gee, I felt bad, and it wasn't much of a Christmas. To tell the truth, I hit the bottle heavy."

11

The Cosa Nostra and traffic in drugs, especially heroin, are almost synonymous. But like everything else about the Cosa Nostra, even this has not always been quite what it seemed. In 1948 Frank Costello, while still acting boss of the Luciano Family, ordered its membership to stay out of dope. According to Valachi, the canny Costello had two reasons for the edict. One was his realization that there were rackets and rackets; bootlegging and gambling, for instance, enjoyed wide public acceptance or, at worst, indifference, while heroin not only was giving organized crime a relative black eye, but also was spurring law enforcement efforts against it in other areas.

Much more important, however, was the dogged harassment of the Bureau of Narcotics. The Cosa Nostra despised—and feared—it, and for the bureau's part, the first feeling was mutual.

Valachi's plaint against the bureau was that it did not play fair, and it is doubtless true that, quite aside from him, some of the bureau's tactics from time to time may have been questionable. A good deal of this is because no other agency has dealt with the Cosa Nostra at such close quarters, known its nature so well or seen so much that it often could do so little about. The Narcotics Bureau, unlike the FBI, which tends to look down its nose at it, does not depend on the informant system for much of its intelligence but regularly engages its agents in dangerous undercover work; they are, as a result, a necessarily somewhat more raffish lot, highly motivated, less disciplined, generally more daring and innovational, occasionally corruptible. Above all else, the Federal Bureau of Narcotics was the first to recognize the existence of an organization like the Cosa Nostra, and no other arm of the law has put more of a crimp in its operations.

Costello's stand produced a varying response in the Cosa Nostra. The Lucchese Family, heavily involved in narcotics, simply ignored it. Others, like the Stefano Magaddino Family in Buffalo, gave it lip service, but privately continued to bring in heroin. The Chicago Cosa Nostra, then led by Tony Accardo, went a step farther than Costello; it is Valachi's understanding that each member who was in dope received $200 a week out of Family funds to help make up for the loss of income.

Even in the Accardo Family, however, the command to get out of narcotics triggered a serious breakdown in Cosa Nostra discipline. The temptation for quick profits was too much, and individual members, particularly those short on cash, persisted in handling heroin secretly. And while Vito Genovese, when he finally displaced Costello, went along with the narcotics ban, it turned out that he had a double standard; if he was given a cut of the take off the top, he always managed to look the other way.

(In those Families, including Genovese's, where heroin was officially "outlawed," a member was on his own if arrested. Cosa Nostra soldiers pay a head tax—during Valachi's day in the Genovese Family it was $25 a month—which is theoretically reserved for such expenses as hiring defense lawyers and supporting wives and children if the upshot is imprisonment. In the 450-member Genovese Family this came to a tidy sum over the course of the year. While a boss in actuality could do whatever he wanted with the money, and often did, a soldier arrested on a narcotics charge knew in advance that he would get no financial help.)

The Bureau of Narcotics had kept an interested eye on Valachi since the mid-1940s, booking him on suspicion in 1944 and again in 1948. "Everyone thought I was in junk," he protests, "but I wasn't. It was because of all them guys that hung around with me. I never made a penny in the junk business at the time."

Then, in early 1956, Valachi was convicted and sentenced to five years in prison in a narcotics conspiracy trial that included his wife's brother, Giacomo (Jack) Reina. Ironically it was the same heroin deal that Eugenio Giannini was arranging when he was murdered. For Valachi it was the first time he had been behind bars since Sing Sing in the 1920s. He did not stay jailed long, however; released on bail pending an appeal, his conviction was reversed on the grounds that the statute of limitations had run out on the part he allegedly played in the conspiracy. It was an extremely complicated case, and admittedly, Valachi's role in it, as charged, was tangential. He says flatly that it was a "bum rap."

If it was guilt by association, as Valachi claims, there was considerable basis for it. As he confessed both to the FBI and in the interviews for this book, two of his codefendants in this case, Pat

Pagano and Pat Moccio, were involved with him in still another heroin deal that apparently escaped detection by the Narcotics Bureau. It took place "around 1952" and was, he says, his first venture in dope. It is also an excellent example of the ethics the Cosa Nostra observes even with its own membership. As Valachi says, "It was a mess. I'm sorry I ever got mixed up in it, and I want the boys who are in it today to know how the greed of the bosses is ruining this thing of ours."

Valachi got the name and address of a French source for heroin from Salvatore (Sally Shields) Shillitani, who had been initiated into the Cosa Nostra at the same time he was. Heroin then could be purchased in France for $2,500 a kilo and sold on the wholesale market in the United States for $11,000. The price, he finally decided, was too attractive to pass up, and with an introduction from Shillitani, he sent Pat Pagano to Marseilles to make contact with the source. Valachi can recall only the source's first name— Dominique.* The method of making contact was a torn dollar bill; Shillitani had already mailed half of it to Dominique, and Pagano went over with the other half. Pagano was successful and returned to inform Valachi that they would be receiving news of a heroin shipment. It was not long in coming:

Pat calls me and says this Dominique's wife is here, and he is going to meet her. He went to meet her at some hotel, I don't know what hotel as it was downtown someplace, and when he finally came back to me, he told me she wants $8,000 for a down payment. I forget if he

---

*Probably Dominique Reissent, a Corsican residing in Marseilles, who, according to the Bureau of Narcotics, imports morphine base into France for conversion into heroin for eventual U.S. distribution.

said how much stuff would be coming. Anyway it was fifteen kilos. If you don't know what a kilo is, it's thirty-five ounces of junk.

Now before I get any deeper, I got to think hard. There is the law about not fooling around with junk, and I got to be careful. I figure the best way to protect myself is to let Tony Bender in on the deal. If Tony goes along, I got nothing to worry about, as he will have to stand by me if I get jammed up. I heard he was in junk. I ain't sure, but I got my suspicions. I figured I would take a chance and talk to Tony. I get heartsick every time I think about it. All I can say is it looked like a good idea. So I spoke to Tony, and I explained to him that I had this proposition. I was a little coy with him, as I was testing him out.

Well, he seemed to be interested, and I told him the rest of it. To make it short, he gave me the $8,000 and another $1,000 expenses for Pat to entertain this Dominique's wife which I asked him for.

I give it to Pat the next day, and now I got problems with him. Pat starts telling me how she seems to like him. I said, "Pat, do me a favor, don't fool around with the man's wife. Do what you have to do, and stick with business." He said, "She is getting bold with me," and I said, "Avoid it."

About four weeks after that we get word that the boat is coming in with the stuff. I'm pretty sure the boat was the *United States*, but don't hold me to it. Now, in the meantime, Sally Shields has got arrested in another case and gets fifteen years, and he's out of it. I tell Tony about the shipment, and he says that he will have Patty Moccio handle everything. It will cost $1,000 a kilo to pay the seaman to get the stuff off the boat, which naturally comes to $15,000. I think it was only one seaman. It don't make no difference. If anybody helps him, he must take care of that out of his end. I don't know how he did it. I'm just happy I don't have to worry about it.

Well, the stuff gets off okay. Now I have to explain that the deal with Tony Bender is fifty-fifty after he gets his $9,000 back, and we pay the rest of what we owe to Dominique. We owe him $29,500, but who cares, as the market for the stuff here runs to $165,000?

This is where that dog Tony comes in. When the stuff was on the way, he was all peaches and cream. After he gets his hands on it, everything changes. He sends for me and says, first of all, that Vito Genovese is in to Frank Costello for $20,000. He don't say what for, it was a loan, and he says, "You know, we will look good with the old man if we take it off the top and pay the debt for him." Then he says that he has brought in some other guys so they can make some money, and he mentions Vinnie Mauro, who is his pet, and Sandino, who is his counselor, and John the Bug, right name Stoppelli, and naturally Patty Moccio, and I don't remember who else. It don't matter, as I will explain later.

In other words, I find I got eight partners counting Pat Pagano. What can I do when he hits me with this? He is my own lieutenant, and that is bad enough. But I can't explain even to Pat, who ain't a member yet—just proposed—that Tony has us because we ain't supposed to be fooling around with junk, and I can't make a beef.

Now with all the deductions, including the $20,000 for Vito, there is $91,500 left. I look at Tony, and I said, "Just give me what's due to me and Pat," and he sends Patty Moccio around to ask do we want it in goods or cash. Well, I talk it over with Pat Pagano, and as the two kids—meaning his brother Joe and my nephew Fiore, we called them the kids—were bothering us for some stuff, I said we would take it in goods, and we got two kilos.

Maybe a couple of months after that, I get a call to go to a certain house in Yonkers. It was Dom the Sailor's house, right name DeQuatro. I went, and who do you think was there? It was Vito Genovese. He said to me, "Did you ever deal in junk?"

I said, "Yes."

He said, "You know you ain't supposed to fool with it."

Vito looked at me and said, "Well, don't do it again."

"Okay," I said.

Of course, I don't pay any attention to this. That is how Vito was. He was just letting me know what a big favor he done me.

Now here's the payoff. A while later I am speaking to John the Bug, and I said, "How did you make out with the money?" and he said, "What money?" I said, "You didn't get any money? You were down as a partner."

He said, "I am down for what, what partner?"

When I realized he didn't know anything about it, I didn't go no further. I didn't want it going back to Tony Bender, as Tony would say, "What are you doing, investigating me?" So I dropped the subject.

A long time after that, in 1956 which is the junk case I beat, I am with Patty Moccio in West Street.* As I never figured Patty got nothing out of the deal, the subject comes up about John the Bug, and to my amazement Patty said, "At least you got the two kilos."

I said, "Are you trying to tell me you got no money?" and he just threw up his right hand, meaning he didn't.

"Oh, my God!" I said. I nearly died.

He said, "Tony was using the oil on me. They need the money for this, they need the money for that, and by the time I went to get my money, there was none left."

I said, "Are you kidding?"

He threw up his hand again and said nothing.

So now you know what happened. There wasn't no nine partners. Mr. Tony Bender and Mr. Vito Genovese just split up the money among themselves.

---

*The Federal House of Detention, 427 West Street, New York City.

On the evening of May 2, 1957, Francesco Castiglia, better known as Frank Costello, dined at the fashionable Manhattan restaurant L'Aiglon. At about 11 P.M. Costello arrived by cab at the apartment building he lived in on Central Park West. As he walked across the lobby, a voice said, "This is for you, Frank." Costello turned just as a shot was fired. Shortly afterward, he arrived at nearby Roosevelt Hospital, blood streaming down his face. He was not, however, seriously wounded. The gunman in his haste had aimed badly, and the bullet only creased Costello's skull.

It was the culmination of Genovese's drive against Costello that had been launched with the murder of Willie Moretti, and it signaled the start of a period of turmoil inside the Cosa Nostra that had not been experienced since the Castellammarese War. While Genovese had already shouldered Costello aside as Family boss, he wanted to consolidate his power, and beyond that he yearned for something more—the mantle of *Capo di tutti Capi*, or Boss of all Bosses, which had gone unworn since the heyday of Charley Lucky Luciano. Symbolically, if not in fact, Costello stood in the way, the czar of a huge gambling empire independent of the Family, his reputation intact, his counsel constantly sought, still held, in that ultimate Cosa Nostra accolade, in high "respect." For Genovese the situation was intolerable; Costello had to be got rid of once and for all.

In the shock waves that followed, the identity of Costello's would-be killer became known not only to just about every member of the Cosa Nostra, but also to the New York Police Department. He was Vincente (The Chin) Gigante, a hulking ex-fighter, now a soldier in the Genovese Family. In commenting professionally on the bungled assassination, Valachi dryly observed, "The Chin wasted a whole month practicing."

With Costello still alive, Genovese had to face the likelihood of a "comeback." His main fear was Albert Anastasia. The "Mad Hatter" and the "Prime Minister," as Costello was sometimes called, were an odd but, nonetheless, close couple. Thus the day after the shooting, according to Valachi, key members of Tony Bender's *regime*, or crew, were called to a hotel on Manhattan's West Side. Approximately thirty soldiers, they were assigned various areas of the city to cover in the event of retaliation. "We were told," Valachi says, "we got to get ready. There could be war over this." Valachi himself was put in charge of East Harlem with five men under him.

Genovese, meanwhile, took to his Atlantic Highlands home with some forty men around him for protection. There he also summoned all the Family lieutenants for a show of loyalty. Only one failed to appear, the aging Anthony (Little Augie Pisano) Carfano, and he would live to rue it. For the moment, however, Genovese wanted a demonstration of complete unity. Tony Bender, who knew Carfano best, was sent to bring him in. "Vito," says Valachi, "told Tony to get him, or else he, Tony, would be wearing a black tie. Well, he did. It wasn't that Little Augie was going to go with Frank. He was just scared. He went back a long ways with Frank, and he thought he was going to get dumped on account of that."

Word of what occurred at the meeting of lieutenants gradually filtered through the Family ranks. Genovese blandly declared that he had been forced to take such drastic action because Costello was actually plotting to kill *him*. He also stated that Costello could no longer be allowed to wield any influence because of his total disinterest in the welfare of Family members and from that point on, anyone caught contacting Costello would have to answer personally to him. Finally he officially confirmed himself head of the Family, in case there were any doubts, and appointed New Jersey–based Gerardo (Jerry) Catena, who had succeeded Willie Moretti, as his underboss. "Who," Valachi notes, "was going to argue?"

While this was happening, the New York police quickly established that Gigante was the probable gunman. He had vanished, however, and at first it was believed that he had been eliminated for botching his assignment. But no. "The Chin was just taken somewhere up in the country to lose some weight," Valachi told me. "I'd say he was around 300 pounds, and you couldn't miss him. They found out that the doorman at Frank's place was half-blind, and they wanted to slim The Chin down, so he, the doorman, won't recognize him."

The retaliation Genovese feared never materialized, and Gigante eventually gave himself up, claiming he had no idea that he had been wanted. Costello, called by the prosecution, maintained that he had not seen the man who shot him and could not imagine who might want to do him harm. This left the elderly doorman as the lone witness, and Gigante's defense attorney had little trouble demonstrating his faulty eyesight.

Costello's docile reaction settled the matter, and even though he had survived the shooting, he was effectively removed as a Genovese rival. It also brought him back into the limelight when he least needed it. At the time he was appealing an income tax evasion conviction, and a slip of paper found in his coat pocket by a detective while Costello was being treated for his wound at Roosevelt Hospital triggered some embarrassing headlines. The slip indicated cash "wins" of $661,284 for an indeterminate period from his gambling casino interests outside New Orleans and in Las Vegas. He refused to answer any questions about this money, wound up in jail with a contempt of court citation, and ultimately went to prison as well on the tax evasion charges.*

Within weeks after the Costello incident, another upper-echelon shooting, this one successful, set the Cosa Nostra on its ear for entirely different reasons.

There has been a great deal of speculative nonsense written about the "ritualistic" aspects of a Cosa Nostra execution of one of its members: that no matter what the offense is, the victim must be killed suddenly and unexpectedly; that he must be wined and dined lavishly before he is disposed of; and that, adhering to an old Sicilian tradition, a shotgun must be used whenever possible. "Naturally," Valachi told me, "you don't want to let the guy know that he's going to be hit, or he might hit you. I never heard of them other things. You just try to be careful and do the best you can."

Thus on the afternoon of June 17, 1957, as he paused for a moment in front of a Bronx fruit store, the end came for sixty-three-year-old Frank (Don Cheech) Scalice, identified by Valachi as the number two man in the Albert Anastasia Family. Four bullets smashed through his neck and head, the New York police report reads, "fired by two unknown white males who fled in an automobile." As usual, nobody could be found who saw anything. The Narcotics Bureau theorized that Scalice, who, among other things, controlled the Bronx construction rackets with an iron hand, had been condemned for failing to deliver on a shipment of heroin allegedly financed by some of his associates.

---

*The casino receipts, of course, would not have been discovered if Costello had refrained from going to the hospital. Since the bullet barely nicked him, I asked Costello once during a magazine interview why he had done so. "Well," he replied, "I put my hand up to my head, and I saw all that blood, and I figured it was still in there."

The reason for his murder, as revealed by Valachi, was infinitely more grave. The Cosa Nostra membership "books" — "We call them books," Valachi says, "but there ain't no real books; it's just the expression we use" — had been closed since the early 1930s. After World War II there were a number of "proposed" members, but the books were not officially opened again until 1954.*

Scalice became one of the most active recruiters. Then it was discovered why. He had actually been selling memberships for amounts up to $50,000, and Anastasia ordered his death forthwith. "We were all stunned when the word got out," Valachi says. "None of that kind of stuff was ever pulled in the thirties. Frank Scalice was the one who started commercializing this thing of ours, and there are others doing it now, believe me. They are making a money deal out of it, rather than the way it used to be. Guys will pay because they want the recognition of being mobbed up. In the old days a man had to prove himself to get it. Today if you give a contract to half them guys, they are liable to drop dead, too."

There was a postscript to Scalice's murder. A brother, Joseph Scalice, made rash vows of vengeance, realized his mistake, and

---

*Among those brought in by Valachi were his nephew, Fiore Siano, the Pagano brothers, and Vincent Mauro. Mauro was his favorite. "He used to come around to this poolroom where we hung out on 108th Street and First Avenue," Valachi recalls. "He looked like a nice kid, quiet and all that, so I got close to him, and I used to take him downtown to the Village Inn and the Hollywood, and Tony Bender took a liking to him. Tony called me one night and said, 'Don't poison his mind.' But I tried to wise him up that my outfit stank and that he should go with another crew. Naturally, he don't listen. He never got nothing from Tony, and now he's doing fifteen years for being in the junk business."

went into hiding. Then he made a bigger mistake. Anastasia spread the word that all would be forgiven, and he returned. On September 7, 1957, Joseph Scalice was reported missing by his son and has not since been seen. Valachi says he was slain in the home of an Anastasia lieutenant, Vincent (Jimmy Jerome) Squillante, and his body butchered into disposable sections. Squillante could call on the perfect vehicle to carry off the remains. He controlled the collection of garbage from New York hotels, restaurants, and nightclubs.

As it turned out, it was the last killing that Anastasia would ever decree. Although Frank Costello had seemingly accepted his forced retirement as a Cosa Nostra power, "Albert A." began to talk out loud about leading a fight to reinstate his old friends. An angry Anastasia was enough to make even Genovese nervous. Save for the fact that Genovese usually had a reason for his savagery and Anastasia often did not, the two men had much in common.

Both flouted all the elaborate rules of order set up in the Cosa Nostra whenever it suited their purpose. In 1951, about the time Genovese was carefully plotting the death of Willie Moretti, Anastasia struck more directly. For years he had been underboss in the Family of Vincent Mangano, one of the original bosses named by Maranzano following the Castellammarese War. But Anastasia got tired of his secondary role. Mangano promptly disappeared and eventually was declared legally dead. Just to tie up loose ends, Mangano's brother Philip was removed from the scene, although his body did turn up in a vacant lot in Brooklyn. Despite such an untidy succession, the Cosa Nostra *Commissione*, or ruling council, meekly approved Anastasia as the new Family head.

Now Genovese discovered that Anastasia was secretly meeting with Costello. Valachi says the two racketeers had worked out a code for each rendezvous. Various hotels were assigned numbers. When they wanted to get together, the number of a particular hotel was relayed, along with the digits of the room.

Apprized of this, Genovese contacted an ambitious Anastasia lieutenant, Carlo Gambino, and convinced him that they would both benefit from Anastasia's death. "Without Vito backing him," says Valachi, "Carlo never would have went for it. But he had a good excuse after Albert broke his promise to Joe Scalice, as it made all the fellows look bad who told Joe it was okay to come in. Besides, Albert was losing heavy at the track, he was there every day, and he was abusing people worse than ever on account of that."

Gambino, according to Valachi, took one of his trusted side-kicks, Joseph (Joe Bandy) Biondo, into the conspiracy. And on October 25, 1957, while Anastasia was relaxing under a pile of hot towels in the barbershop of the Park-Sheraton Hotel in Manhattan, his bodyguard conveniently off on an errand, two gunmen walked in, drew pistols, and riddled him with bullets. Thus, with his flanks neatly covered, Genovese could sit back and watch benevolently as Gambino assumed Anastasia's place as a Cosa Nostra boss with Biondo as the Family's new second-in-command.

(The Justice Department had already been given pieces of this story from none other than Tony Anastasio, who, despite the slight difference in spelling, was a brother of the dead boss. Anastasia had put Tony, nominally a Cosa Nostra soldier, in charge of the huge Local 1814 of the International Longshoremen's Association. As a result, Tony was used to having his own way on the

Brooklyn waterfront. Then, with Gambino in the saddle, he abruptly found himself reduced to figurehead status. The pain of his demotion loosened Tony's tongue, but before he could be fully developed as an informant, he died, of natural causes, in 1963.)

Anastasia's murder created a sensation in the press. But in the avalanche of theories about it that followed, no one in the Cosa Nostra was under any illusion about who was behind it. And it touched raw nerves everywhere. What was Vito Genovese up to? More to the point, when and where was he going to stop? All in all, 1957 had not been a vintage year for the brotherhood. It was going to be even less so before it was over.

On November 14, three weeks after Albert Anastasia was laid to rest, the greatest Cosa Nostra conclave in history took place in a country house in the upstate New York hamlet of Apalachin. There had been other gatherings before, but nowhere near as big. Normally a meeting like this would not have entailed such an influx of mobsters—with more than 100 in attendance—but because of the unsettling atmosphere in which it was being held, each boss, wary about his own future, brought along a retinue of bodyguards.

Although the Justice Department had begun to put together a mosaic of what this now famous underworld summit conference was all about, not until Valachi talked was a complete picture available. It was called together by Genovese. That so many chieftains came on such short notice was in itself an impressive demonstration of his authority.

Originally, Valachi says, Genovese wanted to have everyone convene in Chicago, which offered both a central location, since this was a national meeting, and the usual protective coloring of a big city. But Stefano Magaddino, the Buffalo boss, had a better idea. Why not the Apalachin home of one of his lieutenants, an old Castellammarese named Joseph Barbara? The rural setting, he argued, would be ideal to escape sophisticated metropolitan police surveillance, the peaceful country air ideal for soothing frayed feelings. Genovese finally acceded. Barbara, informed of his hostly duties, sent his son out to make hotel and motel reservations in the surrounding area and then placed a rush order for prime steaks from the local Armour Company meat outlet in the nearby city of Binghamton. But the cut and quality specified by Barbara were unavailable in Binghamton, with its population of 80,000, and a special shipment of the desired steaks had to be trucked in from the Armour plant in Chicago.*

The first order of business was to have the ruling council anoint Genovese as a Family boss. This was really going through the motions; since he had become so powerful, nobody was going to dispute him at this stage, but still a certain formality had to be observed to maintain the Cosa Nostra's intricate, if fragile, organizational structure.

Then Genovese had to deal with the charge that he had violated *Commissione* rules in the attempted assassination of Costello. His pitch was the same as before. Costello was planning, with the connivance of Anastasia, to do him in. Nobody was actu-

---

*The Cosa Nostra, if nothing else, likes to eat well. Recently I noticed a number of its members in a Manhattan restaurant where none had been in evidence before. When I mentioned this to the owner, he said, "Oh, I've got a new chef they like. He came out of retirement, and they all showed up. I had a really big party the other night." Curious about what dishes were preferred, I asked, and was told, "They just said to tell him they were here and to keep the food coming, and they would let him know when they had enough." The restaurateur was delighted with his new, well-behaved clientele. There was none of the bother of charge accounts favored by his chic trade: he was paid in crisp $100 bills.

ally prepared to believe this, but Genovese had a much more convincing brief, since the deed was now past history, that Costello's emphasis on his own enterprises, as opposed to general Family well-being, would inevitably destroy the Cosa Nostra as such and that something had to be done about it.

Genovese also wanted the blessing of his fellow chieftains for the murder of Anastasia. Again nobody really cared about the passing of the kill-crazy boss, but they were concerned about the indiscriminate killing of bosses, and they desired to hear personally, and on the record, from Genovese that there would be no more of this.

Next was the accusation that Anastasia shared with Frank Scalice in the proceeds from the sale of Cosa Nostra memberships and that the Scalice execution was simply a cover-up once word got out about what was going on.

This brought up two proposals for which a meeting would have been called in any event. One was to remove from membership, under the threat of immediate death if they ever talked, all new soldiers who had demonstrated their unworthiness. According to Valachi, they were to be declared "useless and unfit." A count of inept executions had been kept since the membership books were opened. "There were," he says, "twenty-seven contracts that ended in complete misses, slight wounds, and bodies being left around in the street." The other decision up for discussion was to stabilize the status quo of the various Families, thus heading off more trouble, by closing the books.

Finally, on the agenda, was what to do about the Bureau of Narcotics. There was some sentiment for murdering known agents. But the prevailing mood was to outlaw drug traffic for all Families, with death the penalty for disobedience; while realistically it would have much the same effect on some elements in the Cosa Nostra as Prohibition had for the nation, it was considered worth trying as a deterrent.

All of this was supposed to be hashed over at Apalachin. None of it, of course, was. As has been amply reported, far from being the ideal spot for such a conference, it proved to be just the opposite. An alert New York state trooper, Sergeant Edgar D. Crosswell, had noted the arrival in the neighborhood of an unusual number of black limousines full of suspicious passengers. Shortly after noon on November 14, he tracked them to the Barbara home, where the visiting underworld royalty had barely had time for brief, informal exchanges before stepping outside in the pleasant weather to the big barbecue pit and the steaks that would be consumed before getting down to work. Sergeant Crosswell, with initially only three men to help him, decided to set up a roadblock to see what would happen, and the Cosa Nostra panicked.

There are between twenty-five and thirty Cosa Nostra Families of varying sizes, with an estimated membership of 5,000 men, covering every section of the United States. The number of Families is inexact because of the difficulty in establishing the precise independence of smaller ones, which may have only 20 or 30 members. They all were represented, directly or indirectly, at Apalachin. When a Barbara underling burst in with the news of the roadblock minutes after it had been put up, some of those in attendance, among them Genovese, decided to try to brazen their way out by car. They were promptly stopped, taken in for questioning, and searched. Others, however, took to the woods in flight. Some of them, their silk suits a bit worse for wear after scampering through the brambles, were also picked up as state police reinforcements arrived. But many got away.

In all, sixty delegates to Apalachin were netted by Crosswell and his men. They were an affluent group; the cash they had with them amounted to more than $300,000. The Justice Department believes that at least fifty of them escaped. A number of those apprehended were tried on a charge of conspiring to obstruct justice by refusing to explain their presence at Apalachin; they were found guilty, fined, and sentenced, but this was reversed in appeal on the ground that merely meeting as they had did not in itself constitute a crime.

(Almost to a man, they explained they were visiting Barbara to cheer him up because he had a bad heart condition. All other queries brought quick invocation of the Fifth Amendment against self-incrimination. There was one notable exception to this. Apalachin was especially embarrassing to John C. Montana, a lieutenant in the Magaddino Family and one of those nabbed while roaming around the countryside. In 1956 Montana, who had a virtual monopoly over the Buffalo taxicab business, was named Man of the Year by the Erie Club, the official social organization of the Buffalo Police Department. So instead of taking the Fifth, Montana's story was that he was driving to New York City when his brakes failed, and he went to Barbara's house to see if he could get them fixed. After Apalachin, according to Valachi, "he asked his boys to stay away from him as they were making him hot." But when Magaddino heard about this, Montana was finished as a lieutenant. "If you ain't got time for your soldiers," Valachi quotes Magaddino as saying, "step down!")

At the time there were observers who bought the idea that it was all just a friendly get-together of old friends with some fancy police records. Even before Valachi, this seems incredible. It certainly made the hitherto-obscure Barbara one of the most popular figures of the day. Distance seemed to be no object. All the way from Southern California came Frank DeSimone, the Los Angeles Cosa Nostra boss. One can't be sure if Barbara enjoyed the company of San Francisco boss James Lanza; Lanza was not arrested at Apalachin, but he was registered the night before in a hotel fifty miles away. While little more than half the delegates were rounded up, they included James Civello, the Dallas boss, and on hand, representing as unlikely a territory as Colorado, was James Colletti. At Apalachin from the South one found Louis Trafficante, Jr., the Florida boss and the overlord of Cosa Nostra gambling interests in pre-Castro Cuba. Among the few from the Midwest who failed to escape was Frank Zito, a Cosa Nostra power in downstate Illinois. The Detroit Cosa Nostra, headed by Joseph Zerilli, was not in evidence, but no matter; it was well represented by Brooklyn boss Joseph Profaci, whose two daughters had married high in the Zerilli organization. Another Brooklyn boss, Joseph Bonanno, who had a thriving secondary operation in Arizona, was also collared; besides Genovese, so was Cleveland boss John Scalish and then Philadelphia boss Joseph Ida. Chicago boss Sam Giancana was believed to have fled successfully through the timberland.

In the world of the Cosa Nostra, which sets such high store on dignity and respect, the sheer humiliation of it all was the worst part. "I'll tell you the reaction of all us soldiers when we heard about the raid," Valachi said. "If soldiers got arrested in a meet like that, you can imagine what the bosses would have done. There they are, running through the woods like rabbits, throwing away money so they won't be caught with a lot of cash, and some of them throwing away guns. So who are they kidding when they say we got to respect them?"

The Justice Department's Organized Crime and Racketeering Section learned that a series of smaller gatherings were subsequently held around the country to put through the Apalachin agenda, including the confirmation of Carlo Gambino as Anastasia's successor. It was all for naught, however, as far as Vito Genovese was concerned. Within a year, as Cosa Nostra jaws dropped, he was indicted, convicted, and sentenced to fifteen years, which he is currently serving, in a narcotics conspiracy case, the first boss since Luciano to be put away.

project and sell the Lido—"Without the license," he recalls bitterly, "it wasn't worth nothing"—in order to finance his appeal.

He had already milked his loan-shark operation to launch the Yonkers restaurant. "I figured," he says, "that I would just concentrate on it and stay away from the mob." Then Matty, his partner in the dress factory, died, and another dependable source of income went by the boards. Any idea of acquiring a new partner disappeared when Valachi discovered that Matty had not been paying their employee withholding taxes, and all the machinery had to be sold at auction to satisfy a government lien. "I was lucky," he told me, "there wasn't a piece of paper around with my name on it. They call us racketeers, and look what a legitimate guy like Matty was doing."

Strapped for money, Valachi followed the fashion of other Cosa Nostra soldiers in the same fix. He went back into narcotics for a quick buck. This time, however, he did not accept any partners. He knew a number of members who were still in heroin, took a percentage of their shipments, and made his own distribution deals with a "couple of colored fellows." To deliver the heroin, he used a "kid" named Ralph Wagner. "He came from somewhere around Fort Lee, New Jersey, I think," Valachi says. "I forget who sent him to me. Anyway this guy said he's a good kid, and I liked him. He was handy to have around; he helped me paint the house. Ralph was dying to get into junk, so I put him to work. He wanted to get in the mob too, but of course, he can't, as he was a mixbreed, meaning he was part Italian and part German."

Valachi is reluctant to talk about his narcotics period. He says he only dealt in "small amounts to get back on my feet." But one of the "colored fellows" he supplied was John Freeman, Sr., who, according to the Bureau of Narcotics, was an important Negro heroin retailer in Harlem. And Valachi was soon able to resume his shylocking and also to devote himself to building up what showed every sign of becoming a lucrative jukebox operation, or route, as it is called.

At the time Joseph Gallo, a soldier in the Profaci Family in Brooklyn, who would later lead a bloody insurrection against his boss, was attempting to strong-arm his way into control of the jukebox and vending-machine racket in New York. Gallo's methods were primitive. You joined his "union" and took him as a business partner, or else. The alternative, at best, was a brutal beating. This, of course, did not apply to anyone in the Cosa Nostra. When Valachi joined, he did so at his own volition. His dues were minimal—$79 every three months. In return he was given new "locations," which the union guaranteed, in addition to the ones he already had. The advantage to Gallo was that Valachi's membership would be influential in organizing independent operators in East Harlem and the Bronx.

Valachi himself used his shylocking to install jukeboxes in various bars and restaurants. "I lend the guy money, and I don't try to collect the loan," he told me. "In other words, the machine stays in the place, or I get my money back. Joe Gallo was nuts using that rough stuff. That's why they started calling him Crazy Joe. Look what it got him."

(On December 21, 1961, Gallo was sentenced to serve from seven to fourteen years in a state penitentiary for attempted extortion.)

Valachi was so preoccupied with recouping his finances that he barely acknowledged the narcotics arrest and conviction of Vito Genovese. He heard about it before the indictment was announced from Thomas (Tommy Ryan) Eboli, who was rising fast as a Genovese favorite. "There's some Spanish guy testifying against the old man," Eboli said, "and we got to find him. His name is Cantaloupes, you know, like the melon. If you hear anything, give me a ring."

"Naturally I told Tommy okay," he says, "but I ain't going to run around looking for this guy. I felt like telling Tommy I didn't see nobody crying over me when I was arrested, so I forgot about it. Let Vito take care of his own worries. I had enough without him."

(Eboli was slightly off on the name of the "Spanish guy," a key witness against Genovese. He was actually Nelson Cantellops. After a long spell of protective custody, he was released at his own request and in 1965 was mysteriously slain in a bar brawl.)

Valachi was, however, greatly saddened by the murder of his old friend John (Johnny Roberts) Robilotto on September 7, 1958. He says that it was a result of Carlo Gambino's power grab in the Anastasia Family:

When I heard the news, I wasn't surprised. I was just hoping it wouldn't happen, but Johnny was close to Albert. After all, Albert put him into the mob. When Albert was hit, I went over to Brooklyn one night to see Johnny. Some kid took me to the place where he was hanging out. It was some club where he had his office, and I talked privately with him. Naturally I don't come right out with it, but what I'm trying to tell Johnny is I hope he ain't thinking about making no comeback or something because of Albert.

He said, "No, don't worry about it. Tony and Vito already spoke to me," and I said, "Good!" and then I talked about my own affairs, which didn't amount to much, just that the jukebox route was coming along, and that was the end of it. I heard later it was true, him and some other boys—fifteen or twenty—were going to pounce on Carlo, but he beat them to it. Well, no matter what, everyone mourned Johnny Roberts.

By early 1959 Valachi says he was ready to get out of heroin completely. His jukeboxes were then bringing in $500 a week; he had developed a contact in a linen supply company who gave him advance notice on new restaurants, and he expected shortly to double his income. Better yet, the worth of a jukebox route had jumped enormously. Its market value previously had been determined by taking the average revenue for a week, multiplying it by twenty, and adding the depreciated cost of the machines; now the multiple was fifty. Besides his loan shark activity, he also joined with another Cosa Nostra soldier in a neat numbers racket scheme. Valachi limited himself to single-action play—bets on one of the three winning digits each day. This meant that there were ten possible winning numbers, from 0 to 9. Everything over $100 placed on each of these numbers was laid off on another numbers bank. Thus he was only responsible for bets totaling $1,000. Since the odds on the winning single-action number were 7 to 1, Valachi and his partner were assured of a daily profit of $300. Working a six-day week, after deducting the salaries of their runners and "ice" to the police, they still split $1,200. As Valachi notes, "I'd say that was pretty good pay, seeing as we ain't taking no chances."

Valachi's bullish feelings did not last long. One May night in 1959 around eight o'clock—"I was just lucky to be at home"—the phone rang. On the other end was the wife of John Freeman, his heroin retailer, frantically whispering an agreed-on code in case the Narcotics Bureau closed in: "I can't meet you. They've seized all our cars."

Valachi drove away from his house moments before the agents arrived. He then went into hiding in a Bronx apartment he maintained for a girlfriend. He had been aware for some time that he was under periodic surveillance and taken extreme care in visiting her. Operating out of there, he did his best to salvage his shylock loans and turned over the supervision of his jukeboxes to Joseph Pagano. When he realized that Pagano was taking a bit too much income for himself, he canceled the arrangement and placed a "kid named Sally," who had been servicing the machines for him, in charge. "Who would have thought Joey would do a thing like that?" Valachi sadly recalls. "Well, this Sally ain't too smart, but he was honest."

After three months he had a falling out with the girl and headed upstate, stopping in the village of Wingdale, near the Connecticut border. He remained there for another month, then bought a trailer and went into Connecticut. Valachi vaguely considered contacting Girolamo (Bobby Doyle) Santucci, who had moved to Hartford years before. He thought better of it, however, bypassed Hartford, and stopped in a trailer camp in Thompsonville.

All things considered, Valachi looks back fondly on his sojourn there. He liked the people—"They ain't like in New York, they say hello; to tell the truth it caught me off-balance until I got used to it"—and he met a local girl shortly after he arrived. He dated her two or three times a week, attending stock-car races, going to the movies, and drinking occasionally in a roadhouse. "Sometimes," he says, "we would just stay in the trailer and I would cook dinner, as I am a pretty good cook if I must say so." One such evening in September the news came over the radio that Little Augie Pisano had been murdered. "Eh," he remembers thinking, "Vito don't never forget. Little Augie didn't come to the meeting after the business with Frank Costello, and now he has got his."

The girl knew Valachi as Charles Charbano. Initially he told her that he was living in the trailer camp because of domestic trouble with his wife. Eventually he had to admit it was more serious than that, and he said that he was in hiding because he had entered the country illegally years ago and there was a deportation order against him. As they sat listening to a description of the Pisano murder, the radio report added that a onetime beauty queen who was with him had also been killed. The girl said to Valachi, "Only animals would do something like that. I hope you don't know those kind of people."

"Of course," he says, "I told her that I don't, and besides, I had to agree with her. They could have passed Little Augie up and taken care of him another time when he was alone."

Valachi's jukebox caretaker, Sally, visited him regularly while he was a fugitive. In the middle of November he told Valachi that Ralph Wagner, who had been indicted in the same case, was becoming increasingly adamant about seeing him. Valachi gave Sally the phone number of a pay station near the camp and said to have Wagner call him at eleven o'clock the following Friday night. "As I'm waiting for the call," he says, "these agents came out of the dark and stuck me up. They walked me back to the trailer and searched it, and then they took me to Hartford."

Wagner, trying for a ligher prison sentence, had turned him in. Although Valachi suspected this at the time, he decided not to do anything about it. "If I do," he says, "I'm in more trouble." He was then taken for his arraignment in Brooklyn, where John Freeman's son had been nabbed for selling three kilos of heroin to an undercover agent of the Narcotics Bureau. Valachi told the court that he had only learned that he was wanted a few days before and, as a matter of fact, was about to make a phone call to find out why when he was arrested. He was, as a result, released on $25,000 bail. Valachi had already notified a bondsman to be on hand and gave him as security a savings-account bankbook with $3,000 in it.

With both the numbers and the shylocking down the drain because of his absence, he immediately set to work rejuvenating his jukebox route while awaiting trial. In the meantime, he had his lawyer start dickering about the kind of sentence he could expect if he pleaded guilty. The lawyer reported back that it looked like seven years, but after a subsequent conference, according to Valachi, this was changed to twelve years. It was at that point he began to think about jumping bail. "If it's seven years," he told me, "I stay. If it's twelve, I run."

His first impulse was to flee to Brazil. To finance his escape, he counted on selling his jukebox operation, then worth around $50,000, plus using whatever cash he had, $20,000 or so. This plan promptly ran into a snag. No one in the Cosa Nostra wanted to purchase the route for fear of offending the Narcotics Bureau. Attempts to sell it to legitimate interests in the jukebox field were equally futile. "These guys on the outside are scared to buy a route from a mob guy," he explains. "They think the mob guy will sell them the route and wait six months and take all his locations back. Of course, that happens. I tried to tell them that I am in trouble, and there ain't no chance of it happening with me, but it was no good."

After one postponement, Valachi's trial was set for February 1960, with every indication, as he puts it, that he would receive a "lot of time." He had discarded the idea, never really formulated, of going to Brazil. Instead, he would cross the border into Canada.

But with the trial only a few days off, he needed time to work out the details. He tried for a second postponement. When he failed, he agreed to plead guilty if he was given a month to "adjust his affairs." This was granted. Then he went to see Vincent Mauro and Frank (Frank the Bug) Caruso, another soldier in the Genovese Family. "I told them," he says, "I was going to jump bail, and they said they had a friend who will help me. So that's how I meet this Albert Agueci. I met him in Maggie's Bar on Lexington Avenue in the Fifties.* Vinnie Mauro was there, and he introduces me. Now Albert Agueci lives in Toronto, but he is with Steve Magaddino as Steve's Family takes in Toronto and Montreal, and he gives me a certain date to check into the Statler-Hilton Hotel in Buffalo."

Next Valachi took a step that still weighs heavily on his conscience. Afraid that the bail bondsman would seize his Yonkers home after he had gone, he sold it at a loss in a quick cash deal. "I wasn't thinking clear," he told me. "I tried to keep my family, my real family, out of all this, and now I was hurting them." With the money he bought another house for his wife in the Bronx, but he says it could not compare with the one he had relinquished.

That done, he gave his faithful hired hand Sally $1,000, reminded him to keep changing the records in the jukeboxes, and hopped a train for Buffalo. He stayed put in his hotel room for a day and a half until Agueci came for him. Agueci told Valachi that arrangements had been made for him to ride with some "local people" who crossed the border frequently. He was also instructed to tell the border patrol, when asked, that he was born in Buffalo and was going to Canada to "have a good time."

Valachi entered Canada without incident and, after a few miles,

---

*715 Lexington Avenue, near 57th Street. It has since closed down.

transferred to a following car driven by Agueci. They stopped overnight at a motel and reached Toronto the next day. Agueci first took him to his own home, introduced him to his wife and children and his brother Vito, also a Cosa Nostra soldier. He then brought him to the house of an Italian family where he was going to live; Valachi's quarters had a separate entrance and consisted of a kitchen and a "parlor" with a collapsible bed. In the afternoon he went shopping in the neighborhood and bought food and cooking utensils. That evening, having forgotten to get a television set, he decided to look over downtown Toronto, found to his delight that it was reminiscent of Broadway, took in a burlesque show, and returned to the house quite late.

Agueci was waiting for him. Telephone calls from New York had been coming in all night at his home for Valachi. The next one was due at 2 A.M. The caller was Tony Bender himself. "Get back here," Bender said. "The fix is in. You're only going to get five years."

Valachi was due in court for sentencing that day. On the phone he simply replied, "I'll think it over," but before going to bed, he told Agueci that he would not return. Less than an hour later Agueci was back again. "It's out of my hands," he said. "You got to go right back on a plane this morning. The junk agents are putting on too much heat."

His instructions, after landing, were to go straight to the home of the bail bondsman. Valachi did so, and was greeted by the bondsman's cousin, who said, "He's at the courthouse. He'll phone here for you." When the bondsman called, he told Valachi to have his cousin drive him to court. On the way, at a red light, Valachi suddenly jumped out of the car, grabbed a cab, and went to the Bronx. "I just didn't want to go in," he says.

For approximately a month Valachi moved from place to place in the Bronx and in New Jersey—"The people were friends of mine, not members, and I don't want to say their names"—before he learned that he was in deep difficulty with the Cosa Nostra for refusing to accept the five years. At that he called Bender, who insisted, "The fix was all set. It was only going to cost $5,000."

Finally Valachi gave himself up. He remembers sensing that the atmosphere was "all wrong" when he entered the courtroom. He was right. If indeed any deal had ever been arranged, it was no longer in effect, and he would wind up receiving fifteen years. "Naturally," he says, "the first thing I think about is my appeal."* Sentencing was delayed two weeks. During this period, he says, he was questioned continually by Narcotics agents. In a last-ditch try for leniency, he admits that he did divulge some details about heroin traffic, but nothing about either Cosa Nostra members or the Cosa Nostra. Subsequently Ralph Wagner, who had suddenly elected to jump bail himself, was rearrested and given from eight to twelve years. The two men were then sent to the Atlanta Federal Penitentiary.

In Atlanta, according to Valachi, there were about "ninety mob guys." There was a tangible sign of their presence in the prison yard; along with the usual recreational facilities—a baseball diamond, handball and basketball courts, etc.—a large section was reserved for *boccie*. Genovese was the absolute ruler of this little empire within the prison population, the arbiter of all disputes, the dispenser of all favors. One dared not address him unless permission was first granted. One backed away from him after speaking

---

*Valachi's plan was to change his plea to not guilty in an effort to obtain a new trial.

to him. Since he only deigned to come out of his cell two evenings a week, appointments had to be arranged through intermediaries far in advance.*

Thus Valachi had an immediate problem to contend with in Atlanta when Genovese exhibited an unexpectedly cold shoulder toward him from the day he arrived. He was especially concerned about it because all the other Cosa Nostra inmates took their cue from Genovese. At last Valachi was able to arrange a meeting. To his astonishment, it turned out that Genovese considered him to be "with" Tony Bender. Valachi knew that Bender had been downgraded—while in prison, Genovese had appointed Thomas (Tommy Ryan) Eboli and Gerardo (Gerry) Catena to run Family affairs—but he had not realized how "sore" Vito was. This gave Valachi the chance to deliver a few well-chosen words about Bender. "I'm in Atlanta because of Tony," he practically shouted. "Tony told me to come in and I will only get five years. I come in and I get fifteen!" In the ensuing weeks Genovese continued to question Valachi about Bender. It slowly began to dawn on Valachi—"You got to understand that Vito is always talking in curves; it's tough to figure out his real meaning"—that Genovese believed Bender had gone behind his back on several narcotics deals without giving him a cut of the profits. This was confirmed by Genovese's obvious rage, some three months after Valachi arrived in Atlanta, when Bender's "pet," Vincent Mauro, was arrested in a major heroin case.

Having apparently convinced Genovese that he had no part in

---

*While visiting Atlanta once on another story, I happened to be in the tower overlooking the *boccie* courts when a guard pointed Genovese out to me and said, "It's tough on those fellows with him." "How so?" I asked. "Well," the guard said, "he just plays so bad that it's hard for them to lose."

any of this, Valachi settled down to the routine of prison life. He had been assigned to mess-hall duty which meant awaking at 5 A.M. every day. Since Valachi was slightly diabetic, and theoretically subject to, as he says, "dizzy spells," he managed to wangle his way into the more friendly surroundings of the prison greenhouse. Then he was abruptly pulled out of Atlanta in August 1961, and returned to New York for another narcotics conspiracy trial. This was the case which Valachi claims was a "frame job" and which later led to all his "trouble." Among his codefendants were all the people who had helped him go to Canada—Vincent Mauro, Frank Caruso, and Albert and Vito Agueci.

Of the four only the Agueci brothers were at the Federal Detention House on West Street, Mauro and Caruso having fled. Valachi says he took an immediate dislike to Vito, but got along with Albert:

The first thing Albert says to me is, "Joe, I'm sorry you got involved in this." Well, it's too bad what happened to him. I tried to help him, but some guys just can't take being in the can, and I could see right away Albert Agueci wasn't going to last long. All he talked about was getting out on bail. He kept telling me his wife was raising the money to get him out and how he was going to declare himself if Steve Magaddino don't get his brother out, too—meaning he would tell everyone that Steve, his boss, was in on the deal, which he was.

I said, "Albert, how old are you?"

He said, "Thirty-eight."

I said, "You want my advice? You've been sending out too many messages. Take myself, when I got arrested the first time, I don't call no one. I got a bondsman, and I paid for the bond. Everyone does the same thing. Now you're sending out all these messages, and you ain't getting no response. It's bad sending messages out like that. You're going to get into trouble. You ought to be tickled to death they sent you a lawyer."

But Albert is stubborn. He says he don't care. Well, Albert went out on bail because the wife sold the house and put up $15,000. Well, it ain't but a couple of weeks and we hear on the radio that they found Albert's body in some field. He was burned up. They got a print off a finger, and that's how they identified him.

When Albert got killed, they raised his brother's bail to $50,000. But it don't make no difference. The last thing Vito Agueci wants now is to get out.

(Following Albert Agueci's release on a bail, a wiretap picked up a conversation in Buffalo between two lieutenants in the Magaddino Family, unfortunately inadmissible in court. The two lieutenants discussed how Agueci had gone to Magaddino and threatened him. They then began giggling over the prospect of cutting Agueci up and spoke of taking him to "Mary's farm" to work him over. The FBI, believing Mary's farm to be in the vicinity of Buffalo, concentrated the search there. Actually it was near Rochester, New York, where Albert Agueci's body was found in a field on November 23, 1961. His arms had been bound behind his back with wire, and he had been strangled with a clothesline. He was then doused with gasoline and set on fire. An autopsy report revealed that approximately thirty pounds of flesh had been sliced from Agueci's body while he was still alive.)

The trial finally began with Mauro and Caruso still fugitives. "This," Valachi says, "made me the main guy." He admits knowing the government's chief witness, a "workman" for Mauro named Salvatore Rinaldo. He does not dispute the guilt of his codefendants, but he insists that he had nothing to do with this particular case. "This Rinaldo," he says, "just moved the dates around to fit me in." In December, at any rate, having remained silent throughout the trial, he was found guilty along with eleven others and in February 1962, received twenty years, to run concurrently with the term he was already serving.

Three weeks later Valachi was returned to Atlanta. On the way back he considered how to handle Genovese. "Of course," he told me, "I knew Vito would think Tony Bender was in on this deal and left him out of it, and I'm in the middle again." Valachi remained in the penitentiary for only three days before he was yanked back to New York for what he claims was the opening shot in a war of nerves launched by the Narcotics Bureau. During his brief stay in Atlanta Genovese stayed in his cell, and Valachi did not see him. Before he left, however, he was given an ominous message from Genovese expressing the "hope" that he would be back. There was also something else for him to brood about. By now Ralph Wagner had been transferred into Genovese's cell. "I don't know," he says, "if it's Ralph's idea or Vito's. It don't matter. Vito had to give his okay. Who knows what Ralph is telling him? Anything he thinks the old man wants to hear."

In New York, according to Valachi, he was again questioned intensively by narcotics agents in an effort to "break" him. But despite the threat of still another heroin case against him, he says he refused to talk. Then he received stunning news during one interrogation. An agent told him that Tony Bender had been murdered and that he was next on Genovese's list.

(Anthony Strollo, alias Tony Bender, disappeared from his home on April 8, 1962. Mrs. Strollo said that the last words she spoke to her husband were, "You better put on your topcoat. It's chilly." He replied, "I'm only going out for a few minutes.

Besides, I'm wearing thermal underwear." Bender has never been seen again. His wife said that she was baffled by his strange absence. He was, she insisted, a "kind person who had no enemies; everybody loved him.")

When a radio news broadcast several days later confirmed that Bender was missing and presumed dead, Valachi privately recognized his own peril. But his fury at the Bureau of Narcotics was overwhelming, and he was determined to confront the future on his own. "I didn't want to give them, meaning the junk agents," he says, "the satisfaction that they were right."

Now a new element entered the picture. When he was told that he would be sent back to Atlanta, he discovered he had a traveling companion, Vito Agueci. At the time Valachi did not quite know what to make of it. He had heard during his stay at the detention house on West Street that Agueci, instead of being sent to a federal prison after his conviction, had been taken to the Westchester County Jail, where, as he puts it, they keep "all the rats." As if to confirm this, Agueci seemed extremely nervous to Valachi throughout the trip. "He was always asking me," he says, "what everybody thought of him."

But in Atlanta Agueci was placed in the usual thirty-day quarantine for new prisoners, and Valachi forgot about him. He was much more concerned about Genovese, and to his surprise his boss greeted him cordially and even asked if he wanted to move into his cell.* "I should have known about his tricks," Valachi recalls, "but I figured if he wanted to keep an eye on me, I would

---

*The Bureau of Prisons says that in the interests of morale and harmony, its policy, whenever possible, is to grant inmate requests to be in the same cell unless there is some overriding reason against it.

be watching him, too. So naturally I said if you want me to come, I will come. How could I say no to him?"

There were six occupants in the eight-man cell. "Besides me and Vito," he says, "there was naturally Ralph Wagner, another guy who ain't a member and is friendly to Ralph; there is this cripple Angelo, who runs messages for Vito, and some Chink, a Chinaman, who don't know what the hell's going on." During this period there was one disquieting note. Vincent Mauro and Frank Caruso had been apprehended in Spain and brought back to New York just before Valachi had left.* "What did the other guys in the case tell you?" Genovese asked.

"Do you think them guys are going to tell me anything knowing I'm coming here with you? Now if you want to know what I think, not told—"

"No," Genovese snapped, "I don't want to know what you think; I want to know what you know." Then after a moment he said, "Do they know about Tony?"

"Of course they know. Everybody in West Street knows about Tony."

"How do they feel about it?"

"They don't feel good, that's for sure."

"Well," Genovese said lazily, "it was the best thing that could have happened to Tony. He wouldn't be able to take it like you and I."

It seemed to Valachi that Genovese was about to tell him that Bender had been an informer. But suddenly Genovese cut the conversation short. "Let's forget about this," he said, "and get some sleep."

In the middle of May, Vito Agueci came out of quarantine and,

---

*They eventually pleaded guilty.

Valachi says, became a very busy man around the prison yard. First he saw him with John (Johnny Dio) Dioguardi, in Atlanta for income tax evasion. Next he saw him huddled with Joseph (Joe Beck) DiPalermo, who had been convicted in the same narcotics case as Genovese. Finally he learned from a source of his own that Agueci had had an extended conversation with Genovese himself. "Of course, I want to know what's going on," he says, "but I figure it's best to show I don't care."

It is his contention that Agueci, afraid of being tabbed an informer, told Genovese that Valachi not only was secretly involved in heroin, but was now working for the Bureau of Narcotics, and that the bureau knew perfectly well this was going to happen. The Bureau of Narcotics, while acknowledging that Agueci did talk, maintains that this occurred almost two years after the events described by Valachi.

In any event, the reality of Atlanta was that Genovese became increasingly distant, and other members of the Cosa Nostra began to shun him. Once Valachi found himself totally isolated in the section of the mess hall where "us New Yorkers" ate. "I knew something was in the air," he says, "and that it was very serious. It looks like they are trying to make me crack up, but I ain't going to let it happen. I went to Vito and said, 'How can you let them treat me like a dog?' and he said, 'I'll take care of it,' but he never done nothing. He was behind it all."

Then one afternoon after the workday had ended, DiPalermo offered him a steak sandwich that he said had been smuggled out of the prison kitchen. "Now Joe Beck ain't even been speaking to me," Valachi recalls, "and now he wants to give me a sandwich. Naturally, I figured there was something in it, and I threw it away." Fearful of being poisoned, Valachi began eating only packaged foods he bought in the commissary. He also started avoiding the shower room, a favored spot in prison to corner a victim. "The next thing I know," he recalls, "Johnny Dio says to me I must be working late as I missed the last two shower days. Now Johnny hands out the clean clothes in the shower room. He says to come in tomorrow. He will be there, and he will give me my clothes. Well, tomorrow was Wednesday, and the next shower day ain't until Saturday. I said I'll be there, but of course, I don't go."

A few nights later Genovese delivered his little homily to Valachi about removing one bad apple from the barrel before it spoils the rest and solemnly kissed him. "The kiss of death," as Ralph Wagner warned Valachi at the time.

(Wagner perhaps would have been wiser to have said nothing. He was paroled from the Atlanta penitentiary in October 1967. Two weeks later he disappeared. He was last seen in a Manhattan bar at 4 A.M. on October 19. He told his drinking companions that he was on his way to an important meeting in the Bronx. His car was later found abandoned at Broadway and 72d Street.)

After that, all the pretense stopped. The next evening Valachi was walking near the *boccie* courts when Vito Agueci, playing cards on a bench about a hundred feet away, suddenly hurled a stream of curses at him in Italian. "I can't remember all the names he was yelling," Valachi says. "It don't matter. It was the same thing—that I am a filthy dog and a rat. I took a quick look over there, and I could see a lot of guys around him. If I go after him and them greaseballs close in with a knife, that's the end of Joe Cago. So I pretend I don't hear him and keep walking."

When he returned to his cell, he says that not even Genovese could mask his surprise at seeing him alive. Neither man spoke. Valachi lay on his bunk awake through the night. In the morning he requested solitary confinement. "By going into the hole," he told me, "I'm a dead duck. It's just like walking into a police station on the outside. But what's the difference? I'm a dead duck anyway if I get caught in a crowd. How long can I stay away from these guys?"

In solitary he refused to eat. In his fruitless attempt to see George Gaffney, then chief of the New York office of the Bureau of Narcotics, he says he had no real intention, even then, of talking. "I was just going to stall him along to get out of Atlanta." His last hope was crushed when the letter to his wife through which he intended to invoke the aid of Thomas Lucchese was returned to him by prison officials. It was at this point that he decided that if he was going to die, he would at least take "someone" with him. He would not, however, go after Genovese. "I wanted Vito to live," he says, "so someday he should stand accused for the way he treats his men."

On June 22, 1962, after having been ordered back into the prison proper because he would not explain why he had requested solitary confinement, Valachi moved like a mechanical man in the yard. "You can't understand how I felt," he says. "I don't know if I'm coming or going." When he saw three inmates lurking behind the baseball grandstand seemingly in wait for him, he backed up until he saw a pipe lying on the ground. At that moment another prisoner who he thought was DiPalermo went past him, and all the pent-up rage and frustration and despair within Valachi finally exploded. DiPalermo had been one of his chief tormentors. Just before Valachi had gone into solitary, he had sneered, "Hey, Joe, how do you stand with the old man now?"

Then Valachi picked up the pipe and killed the wrong man.

13

Since Valachi's day two of the Cosa Nostra's grand council, Thomas (Three-finger Brown) Lucchese and Joseph Profaci, have died—of natural causes. The *Commissione* normally consists of from nine to twelve Family bosses. At this writing, according to the best estimate of the Justice Department, they are: Vito Genovese, currently in the Levenworth Federal Penitentiary, but still so feared that no one has as yet dared displace him; Carlo Gambino, Joseph (Joe Bananas) Bonanno, and Joseph Colombo, who succeeded Profaci, all of New York; John Scalish of Cleveland; Joseph Zerilli of Detroit; Salvatore (Sam) Giancana of Chicago; Stefano Magaddino of Buffalo; and Angelo Bruno of Philadelphia. Carlos Marcello of New Orleans, as powerful as any boss on the council, sits in occasionally but seems content to observe its rulings. Raymond Patriarca of New England, who was on the council, has lately fallen into disgrace, having allowed himself to be caught and convicted in a murder conspiracy case.

The Commissione is subject to immediate change. Magaddino is seventy-five; Gambino has pleaded a bad heart condition for so long every time he was supposed to appear in court that now even he is reported to be worried about it; Joseph Zerilli, who likes to think of himself as a pillar of the community in Detroit's exclusive Grosse Pointe suburb, is also getting on in years; and Giancana in Chicago has been so hounded by the FBI that his effectiveness as the Family boss is practically nil at this point.

Besides these fairly normal changes, brought about by nature and the law, there may be other, more abrupt ones. Even after Genovese went to prison, no one rushed to challenge his position as the dominating force in the Cosa Nostra for fear that he might win a reversal on his narcotics conviction.

(He lost. The earliest he will be eligible for release is February 6, 1970. After that he faces a deportation order. This does not necessarily mean he is going to go. Paul DeLucia, also known as Paul [The Waiter] Ricca, who followed Frank Nitti as boss of the old Capone Family for a while, was ordered deported in 1957 and is still here, still fighting it. Genovese faces an infinitely more serious problem in the Cosa Nostra itself. It was a member of his Family who talked and who confirmed once and for all the existence of a national crime cartel, whose revelations not only focused public attention on the Cosa Nostra, but spurred a renewed federal law enforcement campaign against it that continues today. Worse yet for Genovese, it was his baseless charge that Valachi was an informer that ultimately led Valachi to speak out. Now he will have to explain away all this to fellow bosses.)

When it became evident that Genovese would not be sprung,

Joseph Bonanno decided to follow in his footsteps. Bonanno had already expanded from his Brooklyn base and, working out of Phoenix, Arizona, claimed a good chunk of the Southwest as his own. Then he decided to go all-out and issued contracts for at least three bosses, Frank DeSimone in Los Angeles, Buffalo's venerable Magaddino and his fellow chieftain in Brooklyn, Carlo Gambino. The grand council, having gone though this once with Genovese, struck back swiftly. Shortly after midnight on October 15, 1964, Bonanno was kidnapped at gunpoint on a Manhattan street by two members of the Magaddino Family. He managed, however, to talk his way to freedom by promising to parcel out his rackets, abdicate as Family boss, and retire to a leisurely life in Phoenix. He laid low for almost two years and suddenly came back fighting. Currently he is engaged in trying to grab Magaddino's Canadian interests. Meanwhile, much to the Cosa Nostra's distress, he is making headlines regularly in New York and littering its streets with bodies in an effort to quell a rebellion in his own Family, led by members who thought he really had resigned.

Publicity of this sort is abhorrent to the modern Cosa Nostra, which envisions itself a much more businesslike operation and considers such goings-on as an unwelcome echo from the past. This is not to say that the younger generation, kept down by their elders hanging onto power, won't make rash moves. Or that, given the opportunity, it will not revert to type. Elements of the Angelo Bruno Family in Philadelphia practically raped the city of Reading, the self-described "pretzel capital" of the world in the heart of the bucolic Pennsylvania Dutch country. Operating in conjunction with a local underworld figure, most of the municipal administration from the mayor on down was corrupted. As a result, the biggest illegal still since Prohibition was tied into the city water supply, the biggest red-light district on the East Coast was set up, and the biggest dice game east of the Mississippi, within an easy drive of either Philadelphia or New York, was launched. Nothing was done for the city. Industry started leaving; downtown Reading became an eyesore. When murmurs of public discontent grew too loud, mob-controlled "reformers" were promptly whisked on the scene. As the city steadily began to wither, a Justice Department task force noticed that the only sign of civic improvement was new parking meters. The company involved in the installation of these meters had a history of kicking back to municipal governments to get the business. It was this thread that eventually unraveled the whole mess, but until outside aid arrived, the local citizenry was truly helpless.

Corruption of public officials has always been a cornerstone of Cosa Nostra operations. A recent example occurred in New York when Antonio (Tony Ducks) Corallo, a lieutenant in the old Lucchese Family specializing in labor racketeering and shylocking, got his hooks into a member of the mayor's inner circle who had gone to him for cash to cover plunges on Wall Street, and a seemingly endless vista of rigged municipal contracts and crooked real estate deals opened up. In this instance an FBI informer was a party to the scheme, and it was stopped almost before it got started. How many similar cases there are in cities around the country where no informant is handy is, of course, the question.

The Cosa Nostra also has its less news-making, but dependable, rackets. Because of the Bureau of Narcotics, it has finally retreated somewhat in one area; while it still controls much of the importation of heroin, it has increasingly left the wholesale market in this country to the Negro and Puerto Rican underworld. It continues to rake in, however, a huge income from illegal gambling and from "skimming"—taking a slice of the receipts off the top before taxes in any venture with a heavy cash flow—vending machines, cigarette machines, jukebox routes, and, perhaps most lucrative of all, licensed gambling casinos.

The current fashion in the Cosa Nostra is the movement into legitimate businesses. The lever is the new sophistication of the loan-shark racket, second now only to gambling in generating income for its membership. The loan shark still preys on those who can least afford it, the poor, and looks for a quick return from such old reliables as bookmakers and B-girl clip joints.

But the idea today is to use the racket to legitimize mob money. It started in fairly obvious areas. If someone who had borrowed cash to acquire a hatcheck concession fell behind in his payments, he simply found that he had a new partner. Instead of wasting time trying to collect from a saloon-keeper who had defaulted, the saloon-keeper was forced to accept jukeboxes and cigarette machines controlled by the Cosa Nostra. Now the operation is infinitely more far-reaching. In an era when the nation's economy has been rapidly expanding, punctuated by periods of tight money in which even the most legitimate businessman can be desperate for cash, everything is fair game. A Wall Street securities house caught short by a sudden market reversal. A builder trapped in a credit squeeze. A garment manufacturer who guessed wrong on this year's line. In return for its money, more often than not, the Cosa Nostra wants a piece of the business.

Once inside a lawful enterprise, members of the Cosa Nostra usually apply their own extortionist ways. To keep a business going, they will use terror tactics to eliminate competition or arrange for advantageous labor contracts through corrupt union officials. Sometimes they simply decide to loot a captive company.

A classic instance of this occurred which involved Valachi's old protégé Joseph Pagano. The management of a large meat wholesaler in New York made the mistake of borrowing money from the Cosa Nostra. Under the guise of safeguarding the loan, the company was required to accept Pagano as its new president. Once he was installed, his colleagues went to work. In ten days they made $1,300,000. They did it by buying huge quantities of meat and poultry on credit and selling it immediately for cash at below-market prices. Then, when they were through, they blithely ordered the thoroughly cowed firm to declare bankruptcy.

While the Cosa Nostra likes to keep out of sight, and often succeeds, its operations still affect everyone whether or not he is aware of it. When it moves into the waterfront, trucking, a huge air cargo center like Kennedy International Airport, service industries like garbage collection, or a monopoly position in hitherto-legitimate businesses, one thing inevitably happens: the cost goes up, and this cost is passed on to the consumer. Trade unionists, meanwhile, pay a special price every time labor racketeers engineer kickback contracts or, as has been particularly true in the teamsters union, use pension funds to finance dubious construction projects.

The drain on the national economy is so enormous that if the Cosa Nostra's illegal profits were reported, the country could meet its present obligations with a 10 percent tax reduction instead of a 10 percent surcharge increase. And nowhere is its impact more pernicious than in poverty-stricken ghettos, where people play the numbers not as a lark, but in a frantic effort to get money. In New York City alone the Cosa Nostra runs a numbers racket that scoops up around $250,000,000 a year. It takes all this cash out of slum areas and, of course, only puts a fraction of it back.

◆ ◆ ◆

Although the Cosa Nostra remains a going concern today, in one respect it will never be the same. Much of its strength rested on its elusive, shadowy nature. Its mystique of secrecy, which had endured for so many years, made it seem omnipotent to its membership. Individuals might fall, but the organization continued, as silent and as real as night. Because of Joseph Valachi, this mystique has been shattered forever. Since he talked—and lived—others in the Cosa Nostra have been persuaded to speak about it for the first time, but none has yet provided the range of intelligence that he has.

(The most recent is Pasquale Calabrese, a soldier in the Magaddino Family in Buffalo. Calabrese, who has been taken to another part of the country and lives with his wife and children under an assumed name, confirmed what Valachi had to say about the structure of a Cosa Nostra Family. His information has also led to the conviction of several members of the Magaddino Family and a disruption of many of its operations.)

As for why he talked, Valachi once wrote me:

Naturally if I could do my life over again, I would. Who would not? Now I am all alone in this world. As you know, I do not write to my wife and my son as they will have nothing to do with me and I don't blame them.

Vito Genovese is responsible for everything. The boys brought him all these stories about me and he believed them. But when he was scheming against me, I was scheming back. Anyway, he found out that he could not do on the inside what he did on the outside. All I did in Atlanta was try to protect myself.

I hope the American people will benefit by knowing what the mob is like. If I was killed in Atlanta, I would have died branded as a rat anyway without doing anything wrong. So what did I lose?

VALACHI'S FBI ARREST RECORD

Author's Collection

Race	White
Sex	Male
Date of Birth	September 22, 1904
Place of Birth	New York City
Age	58
Height	5'8"
Weight	188 pounds

Build	Stocky
Complexion	Dark
Hair	Gray—wavy in front and heavy growth; parts left side and combs back
Eyes	Brown
Scars or marks	None noted
FBI #	544

The following is the arrest record of JOSEPH VALACHI, FBI Number 544:

CONTRIBUTOR OF FINGERPRINTS	NAME AND NUMBER	ARRESTED OR RECEIVED	CHARGE	DISPOSITION
PD, Jersey City, N.J.	Anthony Sorge #B-25	11-10-21	C.W. (loaded revolver)	$100 fine and costs—prob.
	8-17-23, Jos. Valachi, Bronx, N.Y., att. burglary; 10-23-23, sentenced to Sing Sing and received 10-26-23, as Jos. Valachi, #B-75260, 1 year and 3 months to 2½ years, conf. att. burglary, 3rd degree—paroled 8-20-24, 4-9-25, returned on new sentence, delinquent 4-30-25, re-paroled 5-28-26 to begin serving new sentence, discharged by expiration 10-8-26.			
Sing Sing Pr. Ossining, N.Y.	Joseph Valachi #77100	4-9-25	burg. 3rd deg.	3 yrs. Par. 6-15-28
PD, New York, N.Y.	Joseph Valachi #B-58468	3-14-29	A. and R.	3-15-29 dis.
PD, New York City, N.Y.	Joseph Valachi #58468	9-29-29	att. ext.	10-7-29, dis. Returned to Dept. of Correction, Albany, NY, 7-6-34
PD, New York, N.Y.	Joseph Valachi #B-58468	1-13-36	rob.	

USM, Brooklyn, N.Y.	Joseph Valachi #7896	11-1-44	Narcotics	11-20-45, dism by Judge.
Fed. Bur. of Narcotics, Wash., D.C.	Joseph Valachi #SE-191, Fed. Bur. of Narcotics, N.Y., N.Y.	11-1-44	IRC Narcotics	11-20-45 dism by Judge
PD, Balto., Md.	Joseph Michael Valachi #70-356	3-23-48	Investigation Narcotics	3-23-48 inv & rel on chg of inv
Bu of Narc Wash DC	Joseph Michael Valachi #SE-238	5-20-55	conspiracy to violate Fed Narc laws	4-23-56 sent 5 yrs fine $10,000
USM, NY NY	Joseph Valachi #—	5-20-55	Narc vio conspiracy	
Fed Det Hdqtrs NY NY	Joseph Valachi #H 6302	3-21-56	conspy viol Fed Narc Laws	
USP Atlanta Ga	Joseph Valachi #77320	5-18-56	conspy to sell narc	5 yrs
Narc Bu Wash DC	Joseph Michael Valachi #SE-238	7-3-57	unl sale of heroin & conspiracy to violate Fed Narc Laws	3-19-57 See #SE-238 5-20-55 (sent reversed by appeals Crt acquitted)
PD Hartford Conn	Joseph Valachi #32099-N 65	11-19-59	vio Fed Narc Law	
Fed Det Hdqtrs NY NY	Joseph Valachi #H-13210-NY	11-19-59	E-NY vio of Fed Narc Laws Consp	
USM Brooklyn NY	Joseph Valachi #23811	11-20-59	sale of narc	sent 15 yrs fine $10,000 6-3-60
Bu of Narc Wash DC	Joseph Valachi #NY:E 1219	11-19-59	Fed Narc Laws (conspiracy)	
USM New Haven Conn	Joseph Valachi #4453	11-18-59	18 USC 371	
USP Atlanta Ga	Joseph Valachi #23811	6-17-60	vio Narc Laws (rec, concl & sale—T21, S.174; T.18, S.2 USC)	15 yrs

Index

Racehorses: Valachi as owner of, 155–59, 162–66, 172, 189; playing the horses, 159–62

Racketeering, xi, 57n., 63–64, 97, 126; crusade of Dewey against, 124–25, 126–27, 132, 153

Rao, Joseph, 78, 113, 119–20, 121

Rao, Vincent, 50, 56, 105, 138, 140

Reading, Pennsylvania, corruption in municipal government of, 259–60

Reina, Gaetano, 20, 65, 92; murder of, 65; Valachi's marriage to daughter of, 102–6

Reina, Giacomo (Jack), 218

Reissent, Dominique, 219n., 220

Reles, Abe (Kid Twist): as informer, 154; death of, 155

Restaurant racket, 122, 125

Riccobono, Joseph (Staten Island Joe), 184

Rinaldo, Salvatore, 250

Robilotto, John (Johnny Roberts), 144–47, 162, 195, 196–98; murder of, 240–41

Rock, John, 123

Rock, Joseph, 123

Rodino, Peter W., Jr., xviii

Rosenthal, Jack, xiii, xiv, xviii, xix

Ross, Charles, of Waldorf-Astoria Hotel, 97. See also Luciano, Charley (Lucky)

Runnelli, Steve, 74, 78–79, 86

Runners, in numbers racket, 113–14

Rupolo, Ernest (The Hawk), 131–32, 174–75, 178–79; murder of, 179

Russo, Louis, 96

Rye (Ryan), Tommy. See Eboli, Thomas

Salerno, Fat Tony, 138, 139

San Francisco Cosa Nostra, 239

Sandino (Tony Bender's counselor), 140, 221

Santucci, Girolamo. See Doyle, Bobby

Saturday Evening Post, xii

Saupp, John Joseph, 8

Savannah Club, Greenwich Village, 183

Scalice, Frank (Don Cheech), murder of, 226–28, 232

Scalice, Joseph, murder of, 227–28, 229

Scalish, John, 239, 257

Schultz, Dutch, 88, 98, 110–11, 121–25, 126; beer baron of Prohibition, 110–11, 122, 123; and numbers racket, 110–11, 114, 121, 124, 125; and restaurant racket, 123, 125; indictment, 123; murder of, 125

Schuster, Arnold, murder of, 185

Scoop, Charley, 102

Selvagi, Frank, 10–11, 12

Senate investigating subcommittee, xi, xiv, 23–24, 29, 34

Shapiro, Jacob (Gurrah), 47n., 122

Shapiro (numbers controller), 118–20; murder of, 120

Shillitani, Salvatore (Sally Shields), 73, 74, 76, 78, 87, 219, 220

Shylocking. See Loan-sharking

Siano, Fiore (Fury), 203, 206, 211–12, 221, 227n.; disappearance, 209

Sicilians and Neapolitans, hatred between, in Italian underworld, 55, 60, 64, 98

Siegel, Benjamin (Bugsy), 122

Sing Sing, Valachi in, 47, 52–56; attempt on Valachi's life, 53–54, 175; completes seventh grade in prison school, 55

Single action variation of policy game, 110, 241

Siragusa, Charles, 200

"Skimming," Cosa Nostra's income from, 261

Slot machines, 99–101, 108, 134, 136

109–10, 111–21, 126, 141,
150–51, 166–67, 241, 244;
finances horse room, 126; regret
over departure of Luciano, 128;
in restaurant business, 148, 151,
155, 166, 172–73, 181–88, 191,
237–38; in dress manufacturing,
148–51, 155, 166, 189, 238; going
out with the girls, 151–52; as
owner of racehorses, 155–59,
162–66, 172, 189; playing the
horses, 159–62; economic pressure of World War II on operations, 166; in gas ration stamp
racket in World War II, 167–72;
violation of no-hands rule,
181–88; in juke box racket,
189–90, 239, 241–42, 244; as sub-
urbanite, 190–91; and contract
for murder of Giannini, 199–209;
questioned on murder of The
Gap, 214; and narcotics traffic,
218–22, 237, 238, 241; in linen
supply business, 241; in hiding
from Narcotics agents, 241–43;
arrested and arraigned on narcotics charge, 243–44; jumping
bail, 244–47; giving self up, sentencing, 247; trial and sentence
on second narcotics conspiracy
charges, 249, 250–51; efforts of
Narcotics agents to break, 251;
reasons why he talked, 263
Valachi, Mildred Reina, 19, 65, 102–6,
107, 139–40, 151, 159, 163–64,
190–91, 214, 263
Valachi Papers, The (Maas), story
behind, xi–xxiii
Valenti, Jack, xix
Vending-machine racket, 239, 261
Venezia Restaurant, 47, 50
Vernotico, Gerard, murder of, 129–30
Vigorish, 141
Vinson, Fred M., Jr., xviii
Vollero, Alessandro, 55, 60, 101

Wacky brothers, 138–39
Wagner, Ralph, 238, 243, 247, 251,
253, 255; disappearance, 255
Waiters union, Dutch Schultz's con-
trol of, 122
Walker, James J., 99, 108
Washington Star, xiin.
Wedding party of Valachi and Mildred
Reina, 105–6
Weinberg, Abe (Bo), 123–24; disap-
pearance, 124
West, Fats, 58, 100
Westchester County Jail, 10, 15, 17,
21, 33, 252
World War II, exploitation of, by
Cosa Nostra, 167–72

Yonkers home of Valachi: purchase,
190–91; sale, 245

Zangarra, Anthony (Charley Brush),
186
Zerilli, Joseph, 235, 257, 258
Zito, Frank, 235
Zwillman, Abner (Longy), 98

▟ Perennial

Books by Peter Maas:

THE TERRIBLE HOURS: *The Greatest Submarine Rescue in History*
ISBN 0-06-093277-5 (paperback)

On the eve of World War II, the Squalus, America's newest submarine, plunged into the North Atlantic. Miraculously, thirty-three crew members still survived—their ultimate fate depending upon one man, U.S. Navy officer Charles "Swede" Momsen.

"A suspenseful tale of terror, courage, heroism and American military genius."
—Tom Brokaw

SERPICO
ISBN 0-06-101214-9 (mass market paperback)

For years a culture of corruption had pervaded the New York City Police Department. Into this maelstrom came a man who broke the mold: a working class, Brooklyn-born, Italian cop with long hair, a beard, and a taste for opera and ballet. Most of all, Frank Serpico was a man who couldn't be silenced—or bought.

"[A] raw and moving portrait." —*Chicago Sun-Times*

UNDERBOSS: *Sammy the Bull Gravano's Story of Life in the Mafia*
ISBN 0-06-093096-9 (paperback)

In 1992, the highest-ranking member of the Mafia in America ever to defect broke his oath of silence. Gravano's story brings us into the innermost sanctums of the Cosa Nostra—a secret underworld of power, lust, greed, betrayal, and deception, with the specter of violent death always waiting in the wings.

"An absorbing, intimate, alluring tale of power, greed, and Mob intrigue." —*People*

FATHER AND SON
ISBN 0-06-100020-5 (mass market paperback)

The powerful and moving story of Michael McGuire, a successful New York executive and lonely widower, who is suddenly drawn into an IRA gun-running plot and must immerse himself in a violent and twisted web of intrigue in order to save his only son.

"A powerfully tragic story." —*New York Times*

THE VALACHI PAPERS
ISBN 0-06-050742-X (paperback)

When Joe Valachi decided to violate the Cosa Nostra oath of silence, he told everything. His recollection of over 40 years of associations traces the growth and development of the Mafia in the United States, revealing its membership, operations and methods.

"Littered with bodies and unsolved crimes, betrayals and beatings, oaths, ritual and revenge." —*Newsweek*

Available wherever books are sold, or call 1-800-331-3761 to order.